A Student Handbook for Writing in Biology

Second Edition

A STUDENT HANDBOOK FOR WRITING IN BIOLOGY

SECOND EDITION

Karin Knisely
Bucknell University

SINAUER ASSOCIATES, INC.

W. H. FREEMAN AND COMPANY

Cover photos:

Flower: *Semi-cactus Dahlia*; photo by Andrew Sinauer.
Frog: *orange-eyed tree frog*, Litoria sp.; Parrot: *scarlet macaw*, Ara macao; Tortoise: *Aldabra Island giant tortoise*, Geocholone Gigantea; photos by Gerry Ellis/Digital Vision.

Address editorial correspondence to:
Sinauer Associates, Inc., 23 Plumtree Road, Sunderland, Massachusetts 01375 U.S.A.
www.sinauer.com

Address orders to:
VHPS/W. H. Freeman & Co. Order Department, 16365 James Madison Highway, U.S. Route 15, Gordonsville, VA 22942 U.S.A.
www.whfreeman.com

Orders: 1-888-330-8477

Microsoft Word, Microsoft Excel, and Microsoft PowerPoint are registered trademarks of Microsoft Electronics, Inc. In lieu of appending the trademark symbol to each occurrence, the author and publisher state that these trademarked product names are used in an editorial fashion, to the benefit of the trademark owner, and with no intent to infringe upon the trademarks.

Library of Congress Cataloging-in-Publication Data

Knisely, Karin.
  A student handbook for writing in biology / Karin Knisely.— 2nd ed.
     p. cm.
  Includes bibliographical references.
  ISBN-10: 0-7167-6709-0 (pbk.)
  ISBN-13: 978-0-7167-6709-1
  1. Biology—Authorship—Handbooks, manuals, etc. I. Title.

QH304.K59 2004
808'.06657—dc22

                                        2004025038

Printed in U.S.A.
5 4

To those who go the extra mile in pursuit of excellence

CONTENTS

PREFACE

The goal of this handbook is to provide students with a practical, readable resource for communicating their scientific knowledge according to the conventions in biology. *Writing* in biology requires a variety of skills, which include

- Understanding the scientific method (Chapter 1)
- Locating pertinent sources efficiently (Chapter 2)
- Learning to read scientific journal articles and paraphrasing the main ideas, not only to avoid plagiarism, but also to improve comprehension (Chapter 3)
- Describing scientific work in the appropriate format, tone, and style (Chapter 4)
- Writing multiple drafts of papers and revising for clarity and conciseness (Chapter 5)
- Getting feedback from peer reviewers and instructors in a timely fashion and paying attention to mistakes to avoid repeating them in subsequent papers (Chapter 6)
- Becoming proficient at using word processing, computer graphing, and presentation graphics programs (Appendices 1, 2, and 3).

Besides writing about it, we communicate scientific knowledge through posters and oral presentations. In these forms of communication, however, visuals and the speaker's delivery play a greater role than the written material in conveying the message. Tips on preparing good posters are given in Chapter 7. Guidelines for effective oral presentations (Chapter 8) and step-by-step instructions for preparing oral presentations in MS PowerPoint (Appendix 3) are new to the Second Edition.

Furthermore, the Second Edition has been revised and reorganized to increase user-friendliness. The technology sections (Chapter 2 on biology databases, Appendices 1 and 2 on Microsoft Word and Excel) have been updated. To illustrate basic principles for formatting graphs and tables, visuals have been added to complement the textual description. To help students decide on what kind of visual (graph or table) best summarizes their results, data from actual lab exercises are presented with explanations for possible ways to organize the data. Finally, the First Edition's Appendix 4 on faulty writing has been streamlined and incorporated into Chapters 4 and 5. Examples (along with revisions) illustrate typical kinds of mistakes found in different sections of a laboratory report.

The Second Edition is augmented by ancillary materials on the Sinauer Associates Website (www.sinauer.com/Knisely). A **Biology Lab Report Template** in MS Word format can be downloaded. The prompts and formatting are intended to help students get used to scientific paper content and format. Two checklists, **Biology Laboratory Report Checklist** and **Laboratory Report Mistakes**, may be printed out for use in both the revision and feedback stages. Students and peer reviewers can focus on specific points in order to evaluate lab report drafts critically. Instructors may save time providing feedback by using the Checklist for general content and style considerations and the Mistakes key for more specific feedback. Instead of writing a detailed explanation of an error, the instructor may simply write a number corresponding to an entry in the key; it is then the student's responsibility to consult the key for the number's meaning. Finally, the **Evaluation Form for Oral Presentations** may be printed out for use by both students and instructors. This form contains queries on delivery, organization, content, and effectiveness of visual aids, which are intended to provide the speaker with constructive feedback on the success of his or her presentation and areas that need improvement.

Acknowledgments

Feedback from users of the First Edition guided changes made in the Second Edition. In particular, I would like to thank my students in Introduction to Molecules and Cells and Organismal Biology laboratories for giving me suggestions on how to improve this book. The following comment exemplifies the attitude of many first and second-year biology students toward writing, and makes me smile: "I don't enjoy writing lab reports, but your book was extremely helpful in doing so." My colleagues in the Biology Department at Bucknell continue to contribute both a pleasant work environment and valuable feedback on content and readability. I also appreciate the help of Bucknell's Information Services & Resources staff in answering my research and computer-related questions. Special thanks go to Walter H. Piper (Chapman University) for his thoughtful comments on the use of passive and active voice, Warren Abrahamson (Bucknell University) for helping me clarify the use of trendlines, and Karen Patrias (National Library of Medicine) for her insight on citation and reference styles.

I am grateful to all the professionals at Sinauer Associates and W.H. Freeman who helped with the production of this book, especially Andy Sinauer and Sara Tenney, who provided encouragement, feedback, and organizational input on both the First and Second Editions. Carol Wigg and Chelsea Holabird meticulously guided the book through the editorial and production processes and Joanne Delphia incorporated the visuals in an attractive layout. Jefferson Johnson deserves the credit for the cool cover and Jason Dirks and Parker Morse set up the website for the ancillary materials.

I also express my sincere appreciation to those who contributed to the First Edition: John Woolsey (poster presentations), Lynne Waldman (good student lab report), Joe Parsons (University of Victoria), Randy Wayne (Cornell University), John Byram (W.W. Norton & Co.), Kristy Sprott (The Format Group LLC), Wendy Sera (Baylor University), Elizabeth Bowdan (University of Massachusetts at Amherst), Mary Been (Clovis Community College), and the staff of the Bucknell Writing Center.

My passions outside of teaching biology helped shape the style and content of this book. I have always loved sports and, for the past three years, have helped coach the Central Susquehanna girls' lacrosse team. Just as coaching and teaching go hand in hand, excelling on the playing field and in the classroom both require preparation, discipline, teamwork, and a willingness to do one's best. Through lacrosse, I have had the privilege of meeting players, parents, coaches, and umpires who have helped me become a better teacher and coach.

As a freelance German to English technical translator, I appreciate the challenge of choosing just the right words to communicate information faithfully, in a clear and concise manner, and according to conventions familiar to the reader. I am indebted to the translation profession for setting high standards of quality. While the process of writing in a particular genre is challenging, it is also immensely satisfying: We experience the joy of learning, the pleasure of sharing our knowledge with others, and the creativity involved in putting thoughts into words. Especially rewarding is the enthusiasm we generate when we communicate our knowledge well.

Lastly, I would like to thank my family and friends for their unwavering encouragement and love. I am especially grateful to my parents, Elfriede and Adolph Wegener, who have been excellent role models, and who have instilled in me a love of languages and given me a positive outlook on life. My children, Katrina, Carleton, and Brian, continue to be my pride and joy. I would not be able to pursue all of my passions without the support of my husband, Chuck, whose love and sense of humor sustain me.

KARIN KNISELY
LEWISBURG, PA
OCTOBER 2004

Chapter 1

The Scientific Method

Trying to understand natural phenomena is human nature. We are curious about why things happen the way they do, and we expect to be able to understand these events through careful observation and measurement. This is known as the **scientific method**, and it is the foundation of all knowledge in the biological sciences.

An Introduction to the Scientific Method

The scientific method involves a number of steps:

- Asking questions
- Looking for sources that might help answer the questions
- Developing possible explanations (hypotheses)
- Designing an experiment to test a hypothesis
- Predicting what the outcome of an experiment will be if the hypothesis is correct
- Collecting data
- Organizing data to help interpret the results
- Developing possible explanations for the experimental results
- Revising original hypotheses to take into account new findings
- Designing new experiments to test the new hypotheses (or other experiments to provide further support for old hypotheses)
- Sharing findings with other scientists

Most scientists do not rigidly adhere to this sequence of steps, but it provides a useful starting point for how to conduct a scientific investigation.

Ask a Question

As a biology student, you are probably naturally curious about your environment. You wonder about the hows and whys of things you observe. To apply the scientific method to your questions, however, the phenomena of interest must be sufficiently well defined. The parameters that describe the phenomena must be measurable and controllable. For example, let's say that you made the following observation:

> Dwarf pea plants contain a lower concentration of the hormone gibberellic acid than wild-type pea plants of normal height.

You might state the question in the following terms:

> Does gibberellic acid regulate plant height?

This is a question that can be answered using the scientific method, because the parameters can be controlled and measured. On the other hand, the following question could not be answered easily with the scientific method:

> Will the addition of gibberellic acid increase a plant's sense of well-being?

In this example, "a sense of well-being" is not something that can be measured or controlled.

Look for Answers to Your Question

There is a good chance that other people have already asked the same question. That means that there is a good chance that you may be able to find the answer to your question, if you know where to look. Your textbook, journal articles, and the Internet (see Harnack and Kleppinger, 2001 for how to determine a website's reliability) are all good places to begin finding answers. Curiously, attempts to answer the original question often result in new questions, and unexpected findings lead to new directions in research. By reading other people's work, you may think of a more interesting question, define your question more clearly, or modify your question in some other way.

Turn Your Question into a Hypothesis

As a result of your literature search or conversations with experts, you may now have a tentative answer to your original (or modified) question. Now it is time to develop a hypothesis. A **hypothesis** is a possible explanation for something you have observed. **You must have information before you can propose a hypothesis!** Without information, your hypothesis is nothing more than an uneducated guess. That is why you must look for possible answers before you can turn your question into a hypothesis.

A useful hypothesis is one that can be tested and either supported or negated. A hypothesis can never be proven right, but the evidence gained from your observations and/or measurements can provide support for the hypothesis. Thus, when scientists write papers, they never say, "The results prove that… ." Instead, they write, "The results suggest that…" or "The results provide support for… ."

You might transform your question "Does gibberellic acid regulate plant height?" into the following hypothesis:

> The addition of gibberellic acid to dwarf plants will allow them to grow to the height of normal, wild-type plants.

Design an Experiment to Test Your Hypothesis

It doesn't matter whether your hypothesis is ultimately negated. A hypothesis is simply a starting point for designing an experiment to test your explanation. Designing an experiment requires experience, creativity, and a sense for what is practical. Experience may be gained from reading methods published by others and modifying them to suit your purpose. Creativity involves brainstorming (possibly with others) to view the subject from many different perspectives. A sense for what is practical is required, because your experimental design will be limited by considerations such as availability of equipment and supplies, cost, time, and so on.

When you design an experiment, pay attention to the following points.

Define the variables. There are three kinds of variables: dependent variables, the independent variable, and controlled variables. **Dependent variables** are variables such as growth, number of seeds, number of body segments, number of offspring, and so on that you can measure or observe. These variables may be altered by the experimental conditions.

The *one* variable that a scientist manipulates in a given experiment is called the **independent variable**. In a different experiment, a different independent variable can be manipulated. It is important to have *only one* independent variable in a given experiment, because otherwise you don't know which factor is affecting the dependent variable(s).

There are many variables that could possibly affect the outcome of an experiment. Thus, it is important to vary only one (the independent variable) and to keep all others constant. The ones that remain constant are called the **controlled variables**. If, for example, you decide to investigate the effect of gibberellic acid on plant growth, variables such as temperature, humidity, age of the plants, day length, amount of fertilizer, watering regime, and so on would all have to be kept constant. These are the controlled variables.

Design the procedure.   Determine how you will carry out the experiment. The procedure can be based on articles in scientific journals, suggestions from colleagues, intuition or experience, or simply the desire to try out a new idea. The procedure must include the following components.

- **Control treatment.** This is the condition that serves as a reference for the test treatments. Experiments may have both positive and negative control treatments, as illustrated by the following example:
    Hypothesis: The addition of gibberellic acid to dwarf plants will allow them to grow to the height of normal, wild-type plants.
    Negative control: Dwarf plants not given gibberellic acid.
    Positive control: Normal plants not given gibberellic acid.

- **Level of treatment for independent variable.** What concentration of gibberellic acid should be tested? We don't want to choose a concentration that is too low, otherwise we might not see any effect. On the other hand, the concentration should not be so high that it is toxic to the plant. The level of treatment is usually based on previous research (journal articles) or preliminary experiments. The level can even be a range that is appropriate for the biological system or organism to be tested.

- **Replication.** A single result is not statistically valid. The experiment must be repeated several times to see whether similar results are found. The results from several trials may be averaged and analyzed using statistical tests.

Make Predictions about the Outcome of Your Experiment

Before you carry out your experiment, you must have some idea of what results to expect if your hypothesis is correct. Making predictions involves **deductive reasoning**, whereby you use general knowledge or experience to predict the outcome of a specific experiment. For the previous example, you might predict the following results:

- **Test treatment:** Dwarf plants should grow taller when they are treated with gibberellic acid.
- **Negative control:** Dwarf plants treated with plain water should be short.
- **Postive control:** Wild-type plants treated with plain water should reach their normal, tall height.

If the outcome of the experiments matches your predictions, then your hypothesis is supported. If not, then your hypothesis is negated, and you may wish to repeat the experiment, double-check your experimental set-up, and/or come up with a different hypothesis.

Predictions are an important tool, because they give you a sense of direction. But there is a danger here, too: Do not let your predictions affect your objectivity. **Do not make your results fit your predictions—** instead, modify your hypothesis to fit your results.

What is learned from a negated hypothesis can be just as valuable as what is learned from a "successful" experiment. The subsequent modifications of the hypothesis and the associated experiments help researchers gain assurance that their explanation of a particular phenomenon is valid.

Collect Data

Once you have planned and set up your experiment, you are ready to make observations or measurements about your test subject. Consistency is very important in this regard. If, for example, you are measuring plant height, you must always measure according to the same criteria. If you measure from the rim of the pot to the shoot tip the first time, you can't measure from the soil to the shoot tip the next time without introducing error into the results. Sometimes it is necessary to modify during the course of your experiment *how* you collect data and *what kinds of data* are important. These modifications may also be included when you design your next experiment.

You should realize that even some of the most elementary questions in biology have taken hundreds of scientists many years to answer. One approach to the problem may seem promising at first, but as data are collected, problems with the method or other complications may become apparent. Although the scientific method is indeed methodical, it also requires imagination and creativity. Successful scientists are not discouraged when their initial hypotheses are discredited. Instead, they are already revising their hypotheses in light of recent discoveries and planning their next experiment. You will not usually get instant gratification from applying the scientific method to a question, but you are sure to be rewarded with unexpected findings, increased patience, and a greater appreciation for the complexity of biological phenomena.

How to handle variability and unexpected results. Variability is a fact of life. If you toss a coin 10 times, you may get 5 heads and 5 tails the first time, 4 heads and 6 tails the second time, and 8 heads and 2 tails the third time. Even though you used the same coin and tossed it the same way, you got different results in different trials.

This same kind of variability is likely to be present in experimental data. Even when you apply the same treatment to different individuals of the same species of plant, the individual plants may respond differently. How do you know if the different results are due to variability among individuals, the imprecision in making measurements, or some other factor?

This is where statistical analysis comes in. Statistics can help you determine how far you can trust the results when you are sampling a subset of a population in which individuals differ. Statistics can help you decide if the different measurements you obtained for replicates are significant or not.

Some common statistical tests and their uses are given in Table 1.1. Consult a good statistics text (see the Bibliography: Samuels and Witner, 1999; Moore, 2000; Utts and Heckard, 2002) for details on the different tests.

What if your results cannot be explained by variability? Although you would like to give yourself the benefit of the doubt, human error is a possibility, especially in introductory biology laboratory exercises. Human error includes failure to follow the procedure, failure to use the equipment properly, failure to prepare solutions correctly, variability when multiple lab partners measure the same thing, and simple arith-

TABLE 1.1 Common statistical tests and their uses

TEST	APPLICATION
Chi square	To compare how closely the observed or measured data compare to the expected results (e.g., for crosses in genetics)
t-test	To compare the means of two groups
ANOVA	To compare the means of three or more groups

metic errors. If you suspect that human error may have influenced your results, it is important to acknowledge its contribution. If you had time to repeat the experiment, you would first try to eliminate sources of human error rather than revise your original hypothesis.

Organize the Data

Raw data must be summarized and organized before you can begin any interpretation. Two ways to organize data are in tables and figures. Tables have rows and columns and are useful for displaying several dependent variables at the same time or when you want to emphasize the numbers themselves rather than the trend shown by the numbers. Figures are graphs, pictures, diagrams, gel photos, X-ray images, and microscope images—any visual that is not a table.

Line graphs and bar graphs are commonly used in data analysis. **Line graphs** show the effect of the independent variable (the one the scientist manipulates) on a selected dependent variable (the one that changes in response to the independent variable). By convention, the independent variable is plotted on the x-axis, and the dependent variable is plotted on the y-axis. Let's say that for 3 weeks, you treated dwarf and normal plant groups as described under "Predictions" (page 5). If you choose to summarize your results in a line graph, plot time on the x-axis, because time is the variable you controlled. Plot stem height on the y-axis, because this is the variable you expect to change over time, depending on the treatment.

Bar graphs allow you to compare individual sets of data when the data are *discontinuous*. In the previous example, if you wanted to compare the final heights of the plants in the different treatment groups, each bar would represent the height of one group of plants at the end of the 3-week period. The data are discontinuous because different treatment groups are being compared.

The best way to display the data depends on the point you are trying to make. For example, if you want to show that the rate of growth is faster in plants treated with gibberellic acid than that in untreated plants, then a line graph would work well. If you want to emphasize that the end effect of treating plants with gibberellic acid is a taller plant, then a bar graph would be suitable. If you want to compare the actual heights of your plants with those in the literature, then a table would be useful. Also keep in mind that a "picture is worth a thousand words," especially when you are comparing physical differences like height, color, and overall appearance of the organisms. Photographs may call your attention to variables that you didn't measure, but that you might want to consider in future experiments. Photographs are also very effective in poster presentations.

The process of summarizing and organizing data initially involves trial and error, especially when you have a lot of data, the results are variable or unexpected, or you do not have a clear understanding of the purpose of the experiment. These difficulties are a normal part of the learning curve; your organizational ability will improve with practice and experience. See the section on "Organizing Your Data" in Chapter 4 as a starting point.

Regardless of how you choose to display the data, you must be honest. Do not exaggerate axes or trends to support your hypothesis when, in truth, the data do not support it. Show variability when necessary. Use statistical methods to reduce huge amounts of data, and be prepared to explain your reasoning.

Keep in mind that there may be *no difference* between the control and the experimental treatments. If there was no difference, say so, and then try to develop possible explanations for these results.

Try to Explain the Results

Once you have organized the data, you are ready to develop possible explanations for the results. You already found sources on the topic when you developed your hypothesis. Return to this material to try to explain your results. Do your results agree with the findings of other researchers? Do you agree with their explanations? If your results do not agree, try to determine why not. Were different methods, organisms, or conditions employed? What were some possible sources of error?

Revise Original Hypotheses to Take New Findings into Account

If the data support the hypothesis, then you should suggest additional experiments to strengthen the hypothesis. If the data do not support the hypothesis, then you should look for "human error" factors first. If these kinds of factors can be ruled out, suggest modifications to the hypothesis. You may also describe ways to test the new hypothesis. Ideally, scientists will thoroughly investigate a question until they are satisfied that they can explain the phenomenon of interest.

Share Findings with Other Scientists

The final phase of the scientific method is communicating your results to other scientists, either at scientific meetings or through a publication in a journal. When you submit a paper to refereed journals, it is read critically by other scientists in your field, and your methods, results, and conclusions are scrutinized. If any errors are discovered, they are corrected before your results are communicated to the scientific community at large.

Poster sessions are an excellent way to share preliminary findings with your colleagues. The emphasis in poster presentations is on the methods and the results. The informal atmosphere promotes the exchange of ideas among scientists with common interests. See Chapter 7 on how to prepare a poster.

Oral presentations are different from both journal articles and poster sessions, because the speaker's delivery plays a critical role in the success of the communication. See Chapter 8 for tips on preparing and delivering an effective oral presentation.

Chapter 2

Finding Primary References

The development of library research skills is an essential part of your training as a biology student. A vast body of literature is available on just about every topic. Finding exactly what you need is the hard part.

In biology, sources are divided broadly into primary and secondary references. **Primary references** are the journal articles, dissertations, technical reports, or conference papers in which a scientist describes his or her original work. Primary references are written for fellow scientists—in other words, for a specialized audience. The objective of a primary reference is to present the essence of a scientist's work in a way that permits readers to duplicate the work for their own purposes and to refute or build on that work.

Secondary references include encyclopedias, textbooks, and articles in popular magazines. Secondary references are based on primary references, but they address a wider, less-specialized audience. In secondary references, there is less emphasis on the methodology and presentation of data. Results and their implications are described in general terms for the benefit of nonspecialist readers.

You will delve into the biological literature when you write laboratory reports, research papers, and other assignments. Although secondary references provide a good starting point for your work, it is important to be able to locate the primary sources on which the secondary sources are based. Only the primary literature provides you with a description of the methodology and the actual experimental results. With this information, you can draw your own conclusions from the author's data.

Although initially it is difficult to read primary literature, it becomes easier with practice, and your persistence will be rewarded with improved critical thinking skills. A further benefit of reading the primary literature is getting to know the scientists who work in a particular subdiscipline. You may discover that you are sufficiently interested in a subdiscipline to pursue graduate work with one or more of the authors of a journal article. Networking is as important in biology as in other fields.

How do you find primary references that are directly relevant to your topic? The fastest and easiest way is to search computer databases using keywords. Most university libraries have licensing agreements with the companies that produce the databases commonly used in biology. These licenses are almost always restrictive, making them available only to faculty, staff, and on-campus students. Thus, if you wish to research a particular topic using databases such as PubMed, BasicBIOSIS (FirstSearch), and the ISI Web of Science, you must be affiliated with a university or other organization that has such a licensing agreement.

Most of this chapter describes how to find primary references using databases. If you do not have access to these databases, however, you can still locate references the old-fashioned way. This method involves building a bibliography from sources cited in textbooks, journal articles, and other literature (see "References Cited in 'Hits,'" p. 14). This method is laborious and time-consuming because you are doing the physical work of the database, but the end result is often the same.

A disadvantage of using databases is that some older, seminal papers may not be indexed. If your assignment requires a thorough search of the literature, you will have to locate these older references the old-fashioned way.

Current Journal Articles

To find current journal articles in biology, use one of the following databases as a starting point. The descriptions of each database are verbatim from Bucknell University's Information Services and Resources website (<http://www.isr.bucknell.edu/Doing_Research/Databases/index.html>), accessed 2004 Aug 23.

- **PubMed**. The National Library of Medicine connection to the MEDLINE database, which indexes the worldwide literature of medicine and life science from 1965 to the present.

- **BasicBIOSIS (FirstSearch)**. Database of articles from core popular and scholarly life science journals from the last five years.

- **ISI Web of Science**. Provides access to Science Citation Index Expanded (1955– present) and Social Sciences Citation Index (1956–present). Science Citation Index Expanded includes bibliographic information and abstracts found in 5700 major journals across 164 disciplines. Social Sciences Citation Index covers 1725 social sciences journals spanning 50 disciplines. Also provides a Cited Reference Search in both indexes that finds articles that have cited a previously published article.

If you need the full text article *fast*, PubMed and BasicBIOSIS provide the opportunity to download some articles as PDF (*portable document format*) files that can be viewed and printed through Adobe Acrobat Reader, which can be downloaded free of charge from <http://www.adobe.com>. If you need to build a large bibliography on a specific topic very quickly, or if you want to search the literature both forward and backward in time, use the ISI Web of Science. Instructions on using the three databases are given in Boxes A, B, and C.

Electronic journals. Many scientific journals are available electronically. Ask the librarian if your library subscribes to any. The reference and citation formats for electronic journal articles are the same as those for print-journal articles, with the Internet information and date accessed given at the end of the full reference (see "Electronic Journal Articles" in Chapter 4).

Box A. To Use PubMed

1. Ask your librarian for the PubMed URL (*uniform resource locator*). The URL begins with <http://www.> and defines a unique address on the World Wide Web.

2. Enter a keyword or keywords separated by the word *and*. Be as specific as possible to narrow down the number of hits.

3. Click the authors of the hits that sound promising. Alternatively, click the icon to the left of the title; this opens a window containing the abstract of the article and further information on the publication.

4. Read the abstract to see if the article pertains to your topic.
   a. If the article is not appropriate, read the next abstract. If necessary, change the keyword(s) and initiate a new search.
   b. If the abstract seems promising, you may want to read the full article. If the full text is available in electronic format, there will be a box containing the journal name just above the title. Click this box to open a window with the full text. If the text is not available electronically, check to see if your library owns a print version of this journal or request a copy of the article through interlibrary loan.

Box B. To Use BasicBIOSIS (FirstSearch)

1. Ask your librarian for the BasicBIOSIS URL.

2. Enter a keyword or keywords separated by the word *and*. The keyword(s) should be as specific as possible to avoid getting hundreds of inappropriate references.

3. You can also check the **Limit to: Full Text** checkbox if you need the full article text (not just the abstract) fast.

4. Article titles will be listed. Click a title that sounds promising.

5. Read the abstract to see if the article pertains to your topic.
   a. If you limited your search to full text, click **View Full Text** in PDF format to get the article.
   b. If you did not limit your search, and the article sounds promising, check **Availability** (just above the authors' names in the **Detailed Records** window) to see if your library owns this journal. Obtain a print version of the full text so that you can determine whether the article really is relevant to your topic.
   c. If the article is not appropriate, read the next abstract. If necessary, change the keyword(s) and initiate a new search.

To Use Other Databases in FirstSearch

Depending on your topic, you may find some of the other FirstSearch databases helpful:

- MEDLINE (medicine and the health sciences)
- AGRICOLA (agriculture, forestry, animal science)
- GenSciIndex (general science, geology, and earth science)

Avoid databases that specialize in science magazine articles, because these articles are not considered to be primary references. Enter a keyword and examine the abstracts as before.

References Cited in "Hits"

Once you have found a highly relevant journal article, browse the References (Literature Cited) at the end of the article for other relevant journal articles. This kind of search is a very effective way to expand your bibliography, because it provides specific information on the topic of interest.

Box C. To Use the ISI Web of Science (Science Citation Index Expanded)

1. Ask your librarian for the ISI Web of Science URL.
2. Click **Log On to Web of Science**.
3. Click **Easy Search**.
4. Select the checkbox for Science Citation Index Expanded (**Sci-Expanded**)–1955 to present.
5. Click **Topic search**.
6. Type in a keyword(s) for your topic. Be as specific as possible to narrow down the number of hits.
7. Click the title of the hits that sound promising. This opens a window containing the abstract of the article and further information on the publication.
8. At the top of the abstract page under the reference, select **Cited References**. This option shows you the references that were cited in the selected article.
9. On the Cited References page, the authors are listed alphabetically. Other useful information, such as the journal, volume, first page of the article, and year of each author's publication is also provided. Click any entry in blue to see the abstract of the cited reference.

Books

Books may be primary or secondary references. Although books such as *Annual Reviews* are considered secondary references, the Literature Cited sections in review articles often are an excellent source of primary references.

The following steps are usually the ones to take when looking for books on a specific topic in an academic library:

1. Look for the Online Catalog on the library's home page.
2. Select **Search**.
3. Enter keyword(s). Be as specific as possible.
4. Book titles will be listed. Click on any titles that sound promising.
5. Write down the call number and note the availability of the book.

6. When you look for the book on the shelf, browse the titles of other books in the vicinity. Because the Library of Congress cataloging system groups books according to topic, you can often find additional sources shelved nearby.

The Internet

The Internet connects computers around the world. This connection allows you to communicate with other computer users on the network through e-mail, instant messages (IM), chat rooms, and the World Wide Web (www, or the Web).

Connecting to the Internet

To get on the Internet, you need a computer, a modem (to connect your computer to a phone line), an Internet service provider (ISP), and browsing software. If you are affiliated with a school, college, or library, the ISP is probably your organization's computer center; if you are making the connection from your home, you need a commercial ISP such as EarthLink or America Online.

Once you have set up an account with an ISP, you can use a browser to find information on the Internet. Sometimes your browser requires plug-ins (additional software) to work efficiently. Instructions for downloading and installing plug-ins are provided in the browser's message windows.

General Navigation

Figure 2.1 shows the menu bar and the toolbars for the most popular graphic browser, Microsoft Internet Explorer (IE). IE held more than 90% of the market share in desktop browsers in 2004, but critics say that IE has been slower than the other leading browser developers, Opera (<www.opera.com>) and Mozilla (<www.mozilla.org>), to improve security features such as pop-up blocking. Mozilla's browser looks similar to that of Netscape Navigator, a popular browser in the 1990s whose future is presently uncertain. You can run more than one browser on your computer, and the three leading browsers all support both Windows and Macintosh.

Browsers have navigation buttons, menus, and toolbars that you can use to search for information on the Internet. Some useful basic commands are described in Figure 2.1. A comprehensive explanation can be found in Harnack and Kleppinger (2001 and URLs contained therein).

Allows you to save your current search location. In a future session, you merely click the desired Favorite in the list to return to its location.

View/Go To lists the
sites visited recently
in the current session.

Figure 2.1   Screen display from Microsoft Internet Explorer, with a description of two of the most useful navigation features

Search Tools

Search tools (also called search engines) enable you to find information on specific topics by typing in keywords. The search tools scan millions of Internet documents for these keywords and then display a list of the documents (with links) in which these keywords appear.

Two of the most useful search tools are AltaVista (<http://www.altavista.com>) and Google (<http://www.google.com>). Both index millions of Web pages and rank the "hits" according to the frequency of occurrence and relevance to the keywords. Metaengines such as Metacrawler (<http://www.go2net.com>) and Ask Jeeves! (<http://ask.com>) are also useful because they allow you to search several Web indexes simultaneously.

For more information on using search tools to find references, see Harnack and Kleppinger (2001).

Evaluate Internet Sources Critically

Although there is an incredible amount of information available on the Internet, much of this information may be unreliable. Whereas journal articles and books have undergone a rigorous review process, information on the Web may not have been checked by any authority other than the owner of the website.

Be particularly wary of companies and organizations who are more interested in trying to sell a product or an idea than in presenting factual information. The ending of the URL address may help you identify the sponsor, as shown in Table 2.1.

TABLE 2.1 Identifying sponsors of sites on the World Wide Web

TYPE OF WEB PAGE	PURPOSE	ENDING OF URL ADDRESS	EXAMPLES
Informational	To present factual information	.edu, .gov	Dictionaries, directories, information about a topic
Business/ marketing	To sell a product	.com	Coca-Cola, Leica
Advocacy	To influence public opinion	.org	Democratic Party, Republican Party
News	To present very current information	.com	CNN, *USA Today*
Personal	To present information about an individual	Variety of endings, but has tilde (~) embedded in the URL	

Source: Widener University (<http://www.widener.edu/Tools_Resources/Libraries/Wolfgram_Memorial_Library/Evaluate_Web_Pages/659/?vobId=1816>) Accessed 2006 Aug 3.

Websites are not considered primary references. When you write a laboratory report or research paper, make sure most of your references are part of the body of published literature, primarily journal articles. You may, however, supplement the literature cited with information retrieved from the Internet.

Chapter 3

READING AND WRITING SCIENTIFIC PAPERS

No matter whether you are a student or are already engaged in a profession, writing is a fact of life. There are many reasons for writing: to express your feelings, to entertain, to communicate information, and to persuade. When you write scientific papers, your primary reasons for writing are to communicate information and to persuade others of the validity of your findings.

Types of Scientific Writing

Scientific writing takes many forms. As an undergraduate biology major, you will be asked to write laboratory reports, answer essay questions on exams, write summaries of journal articles, and do literature surveys on topics of interest. Upperclass students may write a research proposal for honors work, and then complete their project by submitting an honors thesis. Graduate students typically write master's theses and doctoral dissertations and defend their written work with oral presentations. Professors write lectures, letters of recommendation for students, grant proposals, reviews of articles submitted for publication to scientific journals by their colleagues, and evaluations of grant proposals. In business and industry, scientific writing may take the form of progress reports, product descriptions, operating manuals, and sales and marketing material.

Hallmarks of Scientific Writing

What distinguishes scientific writing from other kinds of writing? One difference is the motive. Scientific writing aims to inform rather than to entertain the reader. The reader is typically a fellow scientist who intends to use this information, for example, to learn more about a process or to improve a product.

A second difference is the style. Brevity, a standard format, and proper use of grammar and punctuation are the hallmarks of well-written scientific papers. The authors have something important to communicate, and they want to make sure that others understand the significance of their work. Flowery language and "stream of consciousness" prose are not appropriate in scientific writing because they can obscure the writer's intended meaning.

A third difference between scientific and other types of writing is the tone. Scientific writing is factual and objective. The writer presents information without emotion and without editorializing.

Scientific Paper Format

Scientific papers are descriptions of how the scientific method was used to study a problem. They follow a standard format that allows the reader, first, to determine initial interest in the paper, second, to read a summary of the paper to learn more, and, finally, to read the paper itself for all the details. This format is very convenient, because it allows busy people to scan volumes of information in a relatively short time, then spend more time reading only those papers that truly provide the information they need.

Almost all scientific papers are organized as follows:

- Title
- List of Authors
- Abstract
- Introduction
- Materials and Methods
- Results
- Discussion
- References

The Title is a **short, informative description of the essence of the paper**. It should contain the fewest number of words that accurately convey the content. Readers use the title to determine their initial interest in the paper.

Only the names of **people who played an active role** in designing the experiment, carrying it out, and analyzing the data appear in the List of Authors.

The Abstract is a **summary of the entire paper** in 250 words or less. It contains (1) an introduction (scope and purpose), (2) a short description of the methods, (3) results, and (4) conclusions. There are no literature citations or references to figures in the Abstract.

The **Introduction** concisely states what motivated the study, how it fits into the existing body of knowledge, and the objectives of the work. The Introduction consists of two primary parts:

1. **Background or historical perspective on the topic.** Primary journal articles and review articles, rather than secondary sources such as textbooks and newspaper articles, are cited to provide the reader with direct access to the original work. Inconsistencies, unanswered questions, or new questions that resulted from previous work set the stage for the present study.

2. **Statement of objectives of the work.** What were the goals of the present study?

The **Materials and Methods** section describes, in full sentences and well-developed paragraphs, **how the experiment was done.** The author provides sufficient detail to allow another scientist to repeat the experiment. Volume, mass, concentration, growth conditions, temperature, pH, type of microscopy, statistical analyses, and sampling techniques are critical pieces of information that must be included. When and where the work was carried out is important if the study was done in the field (in nature), but is not included if the study was done in a laboratory. Conventional labware and laboratory techniques that are common knowledge (familiar to the audience) are not explained. In some instances, it is appropriate to use references to describe methods.

The **Results** section is **where the findings of the experiment are summarized,** without giving any explanations as to their significance (the "whys" are reserved for the Discussion section). A good Results section has two components:

- A *text,* which forms the body of this section
- Some form of *visual* that helps the reader comprehend the data and get the message faster than from reading a lengthy description

In the **Discussion** section, the **results are interpreted** and possible explanations are given. The author may:

- Summarize the results in a way that supports the conclusions
- Describe how the results relate to existing knowledge (literature sources)
- Describe inconsistencies in the data. This is preferable to concealing an anomalous result.
- Discuss possible sources of error
- Describe future extensions of the current work

References list the **outside sources** the authors consulted in preparing the paper. No one has time to return to a state of zero knowledge and rediscover known mechanisms and relationships. That is why scientists rely so heavily on information published by their colleagues. References are typically cited in the Introduction and Discussion sections of a scientific paper, and the procedures given in Materials and Methods are often modifications of those in previous work.

Styles for Documenting References

The Council of Science Editors (CBE Manual, 1994) recommends the following two formats for documenting references:

Citation-Sequence System. In the *text*, the source of the cited information is provided in an abbreviated form as a number in square brackets or parentheses. On the *references pages* that follow the Discussion section, the sources are listed in **numerical order** and include the full reference.

Name-Year System. In the *text*, the source is given in the form of author(s) and year. On the *references pages* that follow the Discussion section, the references are listed in **alphabetical order** according to the first author's last name.

The name-year system has the advantage that people working in the field will know the literature and, on seeing the authors' names, will understand the reference without having to check the reference list. This system is more commonly used and generally is preferred. With the citation-sequence system, for each reference the reader must turn to the reference list at the end of the paper to gain the same information.

Strategies for Reading Journal Articles

Papers in scientific journals are written by experts in the field. Because you are not yet an expert, you will probably find it difficult to read and understand journal articles. The following strategy may help.

Determine the topic. First, try to determine the topic of the article by reading the title and the abstract. Are the authors trying to answer a specific question, explain observations, present a theoretical model of a process, determine the relationship between one or more variables, or accomplish something else?

Acquire background information on the topic. Read about the topic in your textbook. Because textbook authors generally write for a student audience, not a group of experts, your textbook is likely to be easier to read. See "Strategies for Reading your Textbook" for some ways to read biology textbooks efficiently.

Read the Introduction. The introduction is usually easier to follow than the abstract. Skim the introduction with the following questions in mind:

- Why did the author(s) carry out this work?
- What are the main hypotheses?
- What was previously known about the topic or problem?
- What are the objectives of the current work?

Read the Results section selectively. Look at the figures and tables to determine what variables were studied. The independent variable (the one the investigator manipulated) is plotted on the x-axis, and the dependent variable(s) (the one that changes depending on the independent variable) is plotted on the y-axis. Also look for variables in column headings of tables.

There are two places to look for a qualitative description of each figure and table: the figure/table caption, and in the body of the Results section (text). The caption states the main idea of the visual. The topic sentence of the paragraph in the text does the same. Subsequent sentences in the paragraph provide details on what trends or findings the reader should notice in each visual. When you read about the results, ask yourself the following questions:

- What were the independent, dependent, and controlled variables?
- Was there a difference between the controls and the experimental groups?
- What were the main findings regarding the independent and dependent variables?

If necessary, reread the introduction to recall the main objectives and hypotheses of the work. Try to understand the big picture before concerning yourself with the details.

Read the Discussion section. The author typically presents his/her conclusions in this section and describes how the results of the study support these conclusions. This is where you can find out what was learned from the work. In particular:

- Were the hypotheses supported?

- What were the important findings?
- Were there any surprises?
- What further work is necessary or already in progress?
- How does this paper relate to your own work?

Skim the Materials and Methods section. Scan the subheadings (if present) and the topic sentence of each paragraph to identify the basic approach. Do not be concerned with the details at this stage.

Read the article several times. Even experts must read journal articles several times before they understand the methodology and the implications of the findings. Take notes the first time you read the article, noting what is confusing. Consult your notes when you read the article again, and try to clarify what you didn't understand the first time through. Each time you read the article, you will understand a little more.

Strategies for Reading your Textbook

The following strategies are based on the proposition that you cannot read a chapter in a biology textbook just once and understand it completely. Repetition is a key ingredient in learning the material. Repetition not only provides you with multiple opportunities to be exposed to the material, but also gives you time to digest it. The basic approach is to read for organization and key concepts first, and then to fill in the details with each subsequent reading.

The two strategies described here work best with a chapter or section of text no longer than 25–30 pages. The first strategy is proposed by Counselling Services at the University of Victoria, Canada (Palmer-Stone, 2001).

1. Take no more than 25 minutes to:
   - Read the chapter title, introduction, and summary (at the end of the chapter, if present)
   - Read the headings and subheadings
   - Read the chapter title, introduction, summary, headings, and subheadings again
   - Skim the topic sentence of each paragraph (usually the first or second sentence)
   - Skim italicized or boldfaced words

2. Close your textbook. Take a full 30 minutes to:
   - Write down everything you can remember about what you read in the chapter (make a "mind map"). Each time you

come to a dead end, use memory techniques such as associating ideas from your reading to lecture notes or other life experiences; visualizing pages, pictures, or graphs; staring out the window to daydream; letting your mind go blank.
- Figure out how all this material is related. Organize it according to what makes sense in your mind, not necessarily according to how it is organized in the textbook. Write down questions and possible contradictions to check on later.

3. Open your textbook. Fill in the blanks in your mind map with a different colored pencil.

4. Read the chapter again, this time normally. Make another mind map.

A second strategy is:

1. Skim the chapter title, headings, and subheadings for an overview of the chapter content. Write down the headings and subheadings in the form of an outline.

2. Look at your outline and ask yourself the following questions:
- What is the main topic of this chapter?
- How do each of the headings relate to the topic?
- How does each subheading relate to its heading?

3. Read each section, paying special attention to the topic sentence of each paragraph. At the end of each section, summarize the content in your own words. Answer the following questions:
- What's the point?
- What do I understand?
- What is confusing?

4. If you read the assigned pages before the lecture, you can pay attention to the lecture content instead of just frantically taking notes. Check to see if there is a lecture notebook that accompanies your textbook. The lecture notebook contains the figures in black and white, and allows you to take notes during lecture directly on the figures.

5. After the lecture, while the information is still fresh in your mind, reread your notes on your reading. Ask yourself
- What topics did the instructor emphasize in lecture? Fill in your lecture notes with details from your textbook.
- What material do I understand better now?
- What questions remain?

6. Remember that each time you read the material, you will learn a little more.

Study Groups

If you have read the material several times, taken notes, and listened attentively in lecture, but still have questions, talk about the material with your classmates. Small study groups are one reason why students who choose to major in the sciences persist in the sciences, rather than switching to a non-science major (Light, 2001).

What are some benefits of participating in small study groups? One benefit is the comfort level. You may be more likely to talk about problems when you are among your peers; after all, they are not the ones who assign your grade. Secondly, when a group is composed of peers with a similar knowledge base, group members speak the same language. Your instructor speaks a different language, because he or she has already struggled to master the material. When you communicate with your classmates, you verbalize your ideas at a level that is appropriate for your audience of peers. Finally, collaborative learning reflects the way scientists exchange information and share findings in the real world. A spirit of camaraderie develops when people work together toward a common goal. The prospect of learning difficult subject matter is no longer so daunting when you have support from a small group of like-minded individuals. The hard work may even be fun when there is good group chemistry.

Group study is not a substitute for studying alone, however. You must hold yourself accountable for reading the material, taking notes, and figuring out what you do not understand before you meet with your group. If you have not struggled to understand the material yourself, you are not in a position to help a classmate.

Avoid Plagiarism: Paraphrase What You Read

One of the best ways to know whether or not you understand an author's work is to summarize the key findings in your own words. Do not make the mistake of simply copying the text word for word from the textbook or journal article, just because you think that you could not have said it better yourself. Get into the habit of paraphrasing the information in the source document, and carefully noting the source (see "Documenting Sources" in Chapter 4). By convention, direct quotations are not used in scientific writing.

Plagiarism is taking someone else's ideas and passing them off as your own. This includes citing the source but still copying the words verbatim. Usually plagiarism is unintentional and is the result of not having a clear understanding of the material. If you can state in your own words what you think the author meant, then you probably under-stood it. If not, you may have to read the paper a few more times or ask for help from your instructor or a fellow student.

To avoid plagiarism when taking notes, follow this collective advice of Lannon (2000), McMillan (1997), Pechenik (2004), and other authorities on scientific writing:

- Don't take notes until you have read the source text at least twice. You have to understand it before you can summarize it.
- Retain key words.
- Don't use full sentences.
- Use your own words and write in your own style.
- Distinguish your own ideas and questions from those of the source text (e.g., "Me: Confusing!").
- Don't cite out of context. Preserve the author's original meaning.
- Fully document the source for later listing under References.

The Benefits of Learning to Write Scientific Papers

Why is it valuable to learn how to write scientific papers? First, scientific writing is a systematic approach to describing a problem. By writing what you know (and what you do not know) about the problem, it is often possible to identify gaps in your own knowledge.

Secondly, scientific writing aims to persuade the reader of the validity of the procedures, results, and conclusions described in the paper. Improving your reasoning abilities in laboratory reports may have a positive effect on other areas of your life as well.

Third, when you learn to write lab reports, you are investing in your future. Publications in the sciences are affirmation from your colleagues that your work has merit; you have been accepted into the community of experts in your field. Even if your career path is not in the sciences, scientific writing is very logical and organized, characteristics appreciated by busy people everywhere.

Credibility and Reputation

The credibility and reputation of scientists are established primarily by their ability to communicate effectively through their written reports. Poorly written papers, regardless of the importance of the content, may not get published if the reviewers do not understand what the writer intended to say.

You should think about your reputation even as a student. When you write your laboratory reports in an accepted, concise, and accurate manner, your instructor knows that you are serious about your work. Your instructor appreciates not only the time and effort required to understand the subject matter, but also your willingness to write according to the standards of the profession.

Model Papers

Before writing your first laboratory report, go to the library and take a look at some biology journals such as *American Journal of Botany, Ecology, The EMBO Journal, Journal of Biological Chemistry, Journal of Molecular Biology,* and *Marine Biology.* Photocopy one or two journal articles that interest you so you can refer to them for format questions.

Almost all journals devote one page or more to "Instructions to Authors," in which specific information is conveyed regarding length of the manuscript, general format, figures, conventions, references, and so on. Skim this section to get an idea of what journal editors expect from scientists who wish to have their work published.

Because most beginning biology students find journal articles hard to read, a sample student laboratory report is given in Chapter 6. Read the comments in the left margin as you peruse the report to familiarize yourself with the basics of scientific paper format and content, as well as purpose, audience, and tone.

Chapter 4

STEP-BY-STEP INSTRUCTIONS FOR PREPARING A LABORATORY REPORT OR SCIENTIFIC PAPER

In order to prepare a well-written laboratory report according to accepted conventions, the following skills are required:

- A solid command of the English language
- An understanding of the scientific method
- An understanding of scientific concepts and terminology
- Advanced word processing skills
- Knowledge of computer graphing software
- The ability to read and evaluate journal articles
- The ability to search the primary literature efficiently
- The ability to evaluate the reliability of Internet sources.

If you are a first- or second-year college student, it is unlikely that you possess all of these skills when you are asked to write your first laboratory report. Don't worry. The instructions in this chapter will guide you through the steps involved in preparing the first draft of a laboratory report. Revision is addressed in the next chapter, and the Appendices will help you with word processing and graphing tasks.

Timetable

Preparing a laboratory report or scientific paper is hard work. It will take much more time than you expect. Writing the first draft is only the first step. You must also allow time for proofreading and revision. If you work on your paper in stages, the final product will be much better than if you try to do everything at the last minute.

TABLE 4.1 Timetable for writing your laboratory report

TIME FRAME	ACTIVITY	RATIONALE
Day 1	Complete laboratory exercise.	It's fun. Besides, you need data to write about.
Days 2–3	Write first draft of laboratory report.	The lab is still fresh in your mind. You also need time to complete the subsequent tasks before the due date.
Day 4	Proofread and revise first draft (hard copy).	Always take a break after writing the first draft and before revising it. This "distance" gives you objectivity to read your paper critically.
Day 5	Give first draft to a classmate for review.	Your peer reviewer is a sounding board for your writing. He/she will give you feedback on whether what you intended to write actually comes across to the reader. You may wish to alert your peer reviewer to concerns you have about your paper (see "Look at the Big Picture" in Chapter 5).
	Arrange to meet with your classmate after he/she has had time to review your paper ("writing conference").	An informal discussion is useful for providing immediate exchange of ideas and concerns.

The timetable outlined in Table 4.1 breaks the writing process down into stages, based on a one-week time frame. You can adjust the time frame according to your own deadlines.

Format Your Report Correctly

Although content is important, the appearance of your paper is what makes the first impression on the reader. If the pages are out of order and the ink is faded, subconsciously or not, the reader is going to associate a sloppy paper with sloppy science. You cannot afford that kind of reputation. In order for your work to be taken seriously, your paper has to have a professional appearance.

Scientific journals specify the format in their "Instructions to Authors" section. If your instructor has not given you specific instructions, the layout specified in Table 4.2 will give your paper a professional look.

TABLE 4.1 *Continued*

TIME FRAME	ACTIVITY	RATIONALE
Day 6	Peer reviewer reviews laboratory report.	The peer reviewer should review the paper according to two sets of criteria. One is the conventions of scientific writing as described in "Scientific Paper Format," and the other is the set of questions in "Get Feedback" in Chapter 5.
	Hold writing conference during which the reviewer returns the first draft to the writer.	An informal discussion between the writer and the reviewer is useful to give the writer an opportunity to explain what he/she intended to accomplish, and for the reviewer to provide feedback.
Days 6–7	Revise laboratory report.	Based on your discussion with your reviewer, revise as necessary. Remember that you do not have to accept all of the reviewer's suggestions.
Day 8	Hand in both first draft and revised draft to instructor.	Your instructor wants to know what you've learned (we never stop learning either!).

Consult the sample "good" student laboratory report in Chapter 6 for an overview of the style and layout. An electronic file called "Biology Lab Report Template," available at <www.sinauer.com/Knisely> is formatted according to the guidelines of Table 4.2 and provides prompts that help you get started writing in scientific paper format. For details on how to format documents in Microsoft Word, see "Formatting a Document" in Appendix 1.

Computer Savvy

Know your PC and your word processing software. Most of the tasks you will encounter in writing your laboratory report are described in Appendix 1 "Word Processing Basics" and Appendix 2 "Making XY Graphs in Excel." If there is a task that is not covered in these appendices, write it down and ask an expert later. If you run into a major

TABLE 4.2 Instructions to authors of laboratory reports

FEATURE	LAYOUT
Paper	8½" x 11" (or DIN A4) white bond, one side only
Margins	1.25" left and right; 1" top and bottom
Font size	12 pt (points to the inch)
Typeface	Times Roman or another *serif* font. A serif is a small stroke that embellishes the character at the top and bottom. The serifs create a strong horizontal emphasis, which helps the eye scan lines of text more easily.
Symbols	Use word processing software. Do not write symbols in by hand.
Pagination	Arabic number, top right on each page except the first
Justification	Align left/ragged right or Full/even edges
Spacing	Double
New paragraph	Indent 0.5"
Title page (optional)	Title, authors (your name first, lab partner second), class, and date
Headings	Align headings for Abstract, Introduction, Materials and Methods, Results, Discussion, and References on left margin or center them. Use consistent format for capitalization. Do not start each section on a new page unless it works out that way coincidentally.
Subheadings	Use sparingly and maintain consistent format.
Tables and figures	Incorporate into text as close as possible after the paragraph where they are first mentioned. Use descriptive titles, sequential numbering, proper position above or below visual. May be attached on separate pages at end of document, but must still have proper caption.
Sketches	Hand-drawn in pencil or ink. Other specifications as in "Tables and figures," above.
References	Citation-Sequence System: Make a numbered list in order of citation.
	Name-Year System: List references in alphabetical order by the first author's last name. Use a hanging indent (all lines but the first indented) to separate individual references.
	Both systems: Use accepted punctuation and format.
Assembly	Place pages in order, staple top left.

problem that prevents you from using your PC, you should have a backup plan in place (familiarity with another PC).

Always back up your files. Images generally require more storage space than text, which means that large files may not fit on a floppy disk; use a zip disk or CD instead, or use a jump drive to transfer files from one computer to another via the USB port.

Save your file frequently while writing your paper. Set up your computer to automatically save data every 10 minutes, so that in the event of a power failure you will have lost only 10 minutes' worth of work. These tasks are described in Appendix 1.

Install antivirus software on your computer, and always check disks and CDs for viruses before you use them. Beware of files attached to e-mail messages. Do not open attachments unless you are sure they come from a reliable source.

Store floppy disks, zip disks, and CDs in protective boxes. Keep them away from magnetic devices (TV, speakers, etc.) and excess humidity, heat, and cold.

If you must eat and drink near a computer, keep beverages and crumbs away from the hard drive and keyboard.

Getting Started

Set aside 1 hour to begin writing the laboratory report as soon as possible after doing the laboratory exercise. Restricting the time to 1 hour forces you to make effective use of limited time. It also takes advantage of the attention span to which you are accustomed in lecture.

If your paper is progressing well and you are having fun, extend the 1-hour period until you perceive a loss of concentration. Promise yourself a reward after a certain amount of progress. It is not necessary to deprive yourself of pleasure as long as you work efficiently.

Reread the Laboratory Exercise

You cannot begin to write a paper without a sense of purpose. What were the objectives of your experiment? What questions are you supposed to answer? Take notes on the laboratory exercise to prevent problems with plagiarism when you write your laboratory report.

Audience and Tone

Scientific papers are written for scientists. Similarly, laboratory reports are intended to describe procedures and findings to fellow students. Of course, your instructor is going to read and evaluate your laboratory report, but your instructor is not the audience for whom you are writing. You are *writing for your peers*—students just like yourself.

Write as though your peers are fellow scientists, not students in a classroom situation. Note the difference between the original text and the revision in the following examples:

> EXAMPLE 1:   The experiments performed by the students dealt with how different wavelengths of light affect seed germination.

> REVISION: The purpose of the experiment was to determine how different wavelengths of light affect seed germination.

> EXAMPLE 2:   The purpose of this experiment is to become acquainted with new lab techniques such as protein analysis, serial dilutions, and use of the spectrophotometer.

> REVISION: The purpose of this experiment was to use the Biuret assay to determine protein concentration in egg white.

Your audience has a knowledge base similar to your own. Thus, when you introduce your topic and describe and interpret your results, you should assume that your audience will know some scientific vocabulary, but you should clarify or define less familiar terms. When deciding on how much background information to include, assume that your audience knows what you learned in class.

Keep the tone of your laboratory report factual and objective. Do not use jargon (terms known only to experts) or copy information verbatim from journal articles or your textbook. Remember that your objective is to write for your peers in a style that they can understand clearly.

Start with the Materials and Methods Section

The order in which you write the different sections is not the order in which they appear in the finished laboratory report. The rationale for this plan will become obvious as you read on. The Materials and Methods section requires the least amount of thought, because you are primarily restating the procedure in your own words.

Writing Style

Laboratory *exercises* are typically written in present tense, often with instructions in the form of a numbered list. When you write your laboratory *report*, however, summarize what you did in **full sentences and well-developed paragraphs**. Use **past tense**, because you completed the experiment some time ago.

You will have to make decisions on verb voice when you describe the procedure. For example:

ACTIVE VOICE: I peeled and homogenized the potatoes.

PASSIVE VOICE: The potatoes were peeled and homogenized.

The sentence written in active voice is more natural and dynamic, but it shifts the emphasis from the subject, "the potatoes" to "I." When you describe a procedure, *who* did it is not as important as *what* was done. Passive voice places the emphasis on the potatoes, where it belongs. Take care not to overuse passive voice, however, because it can make your writing seem stuffy and impersonal. See the section "Active and Passive Voice" in Chapter 5 for appropriate use of active and passive voice.

Details: To Include or Not To Include?

When you first begin to write scientific papers, it may be hard to decide how much detail to include in the Materials and Methods section. On the one hand, too many details bore the reader; on the other hand, you must include enough information to allow the reader to repeat the experiment. The following are some examples of things to avoid and include in Materials and Methods sections.

Avoid listing materials. The separation of the heading "Materials and Methods" into two parts is understandably confusing for beginning students. The implication is that the materials should be listed separately from the methods. In fact, **materials should not be listed separately** unless the strain of bacteria, vector (plasmid), growth media, or chemicals were obtained from a special or noncommercial source. It will be obvious to the reader what materials are required on reading the methods.

Avoid any reference to the container. Consider the following example:

EXAMPLE: Eight clean beakers were labeled with the following concentrations of hydrogen peroxide and those solutions were created and placed in

> the appropriate beaker: 0, 0.1, 0.2, 0.5, 0.8, 1.0,
> 5.0, and 10.0.

There are two problems with this passage: (1) It is not necessary to mention that clean, suitable containers were used to store the solutions; this is common practice in the laboratory. (2) The contents are more important than the label and the appropriate units must be mentioned.

> REVISION: The following hydrogen peroxide solutions were
> prepared: 0, 0.1, 0.2, 0.5, 0.8, 1.0, 5.0, and 10.0%.

Avoid lengthy descriptions of routine procedures. Your peers have learned certain basic laboratory techniques, such as using an electronic balance to weigh chemicals, a vortex mixer to mix solutions, a spectrophotometer to measure absorbance (optical density) of solutions, and micropipettors to measure small volumes. The amount of detail in the following examples is therefore unnecessary.

> EXAMPLE: To make the dilution, a micropipettor was
> used to release 45, 90, 135, and 180 µL of
> bovine serum albumin (BSA) into four differ-
> ent test tubes. To complete the dilution, 255,
> 210, 165, and 120 µL of TBS was added, respec-
> tively.

> REVISION: The following concentrations of BSA were pre-
> pared for the Bradford assay: 300, 600, 900,
> and 1200 µg/mL.

With appropriate instruction, making dilutions of stock solutions becomes a routine procedure. In the above example, you should assume that your peers can make the solution using the appropriate measuring instruments *as long as you specify the final concentration.*

> EXAMPLE: The test tubes were carried over to the spec-
> trophotometer and the wavelength was set to
> 595 nm (nanometer). The spectrophotometer
> was zeroed using the blank. Each of the
> remaining 8 samples in the test tubes were
> individually placed into the empty spec tube,
> which was then placed in the spectrophotome-
> ter where the absorbance was determined.

> REVISION: The absorbance of each sample was measured
> with a spectrophotometer at 595 nm.

Avoid giving "previews" of your data analysis in the Materials and Methods section. Consider the following passage:

> EXAMPLE: A graph was plotted with Absorbance on the *y*-axis and Protein concentration on the *x*-axis. An equation was found to fit the line, then the unknown protein absorbances that fell on the graph were plugged into the equation, and a concentration was found.

Making graphs is something that you do when you summarize your raw data, but it is not part of the experimental procedure. How and why you chose to organize the data will become obvious to the reader in the Results section, where you display graphs, tables, and other visuals and describe the noteworthy findings.

> REVISION: *Delete this entire passage.*

Include all relevant information needed to repeat the experiment. Consider the following example:

> EXAMPLE: In this lab, we mixed varying amounts of BSA stock solution with varying amounts of TBS using a vortex mixer. We used a spectrophotometer to measure absorbance of the 4 BSA samples, and then we determined the concentration of 4 dilutions of egg white from the standard curve.

This procedure does not give the reader enough information to repeat the experiment, because essential details like *what concentrations of BSA* were used to construct the standard curve, *what dilutions of egg white* were tested, and the *wavelength* set on the spectrophotometer are left out.

> REVISION: BSA solutions (2, 3, 5, 10 mg/mL) were prepared in Tris-buffered saline (TBS). The egg white sample was serially diluted 1/5, 1/15, 1/60, and 1/300 with TBS. The absorbance of all samples was measured at 550 nm using a Spec 20 spectrophotometer.

Cite published sources. If you are paraphrasing a published laboratory exercise, it is necessary to cite the source (see "Documenting Sources"). Unpublished laboratory exercises are not usually cited; ask your instructor to be sure.

Do the Results Section Next

The Results section is a *summary* of the key findings of your experiment. This section has two components:

- Visuals, such as tables and figures
- A body, or text, in which you describe the findings presented in the visuals.

Describe the findings in an objective manner, without explaining why or discussing possible implications.

When you work on the Results section, you will complete the following tasks, which are often done concurrently, not necessarily sequentially:

- **Analyze the raw data.** Raw data are all the observations and measurements that you recorded in your lab notebook. It is your job as the author to analyze all these data and process the information for the reader. **Do not simply transfer raw data into your lab report** (your instructor may ask you to attach pages from your lab notebook as an appendix, however). Instead, summarize the data by eliminating aberrant results (because you realized that you made a mistake in obtaining these results), averaging replicates, using statistical methods to see possible trends, and/or selecting representative pictures (micrographs or gel images). The goal of data analysis in general is to try to figure out what the data show; more specifically, you are interested in whether the data support or negate your hypotheses.

- **Organize summarized data in tables or figures.** When you organize summarized data in a table or plot numerical values on a graph, you may be able to see trends that were not apparent before. Effective visuals are more powerful than words alone and they provide strong support for your arguments. See "Organizing Your Data."

- **Decide in which order to present the tables and figures.** The sequence should be logical, so that the first visual provides a basis for the next or so that the reader can easily follow your line of reasoning.

- **Describe each visual in turn and refer to it parenthetically.** You should first describe what you want the reader to notice about the visual. Refer to the visual by number in parentheses at the end of the first sentence in which you describe it. Position the visual after the descriptive paragraph (Figure 4.1). Other examples of good descriptions of visuals are given in Figures 4.2 and 4.4.

Tips for Writing the Body of the Results Section

Specify the visual that contains the data that you are describing. Refer to Figure 4.1 as you read the following passage:

EXAMPLE: The results that were obtained while completing the lab clearly showed that the percentage of male gametophytes increased with population density initially, but above 75 gametophytes/plate, the percentage of males leveled out at about 72%.

This passage is unsuitable because we don't know where to find the results the author is describing. Always refer to the visual parentheti-

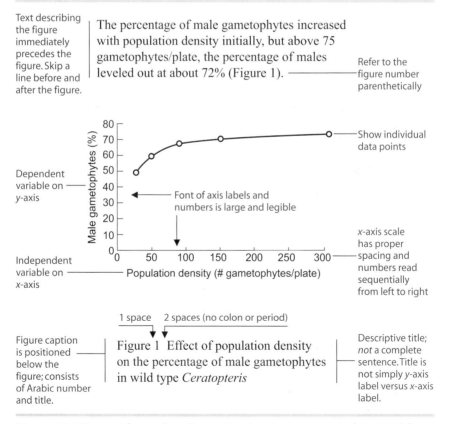

Text describing the figure immediately precedes the figure. Skip a line before and after the figure.

The percentage of male gametophytes increased with population density initially, but above 75 gametophytes/plate, the percentage of males leveled out at about 72% (Figure 1).

Refer to the figure number parenthetically

Show individual data points

Dependent variable on y-axis

Font of axis labels and numbers is large and legible

x-axis scale has proper spacing and numbers read sequentially from left to right

Independent variable on x-axis

Population density (# gametophytes/plate)

1 space   2 spaces (no colon or period)

Figure caption is positioned below the figure; consists of Arabic number and title.

Figure 1 Effect of population density on the percentage of male gametophytes in wild type *Ceratopteris*

Descriptive title; *not* a complete sentence. Title is not simply y-axis label versus x-axis label.

Figure 4.1 Excerpt from a Results section showing a properly formatted figure with one line (data set); text that describes the figure precedes it

cally so that the reader can look at the visual presentation of the data while reading the author's description of those data.

REVISION: The percentage of male gametophytes increased with population density initially, but above 75 gametophytes/plate, the percentage of males leveled out at about 72% (Figure 1).

Eliminate unnecessary introductions. Look at Figure 4.2 as you read the following example:

EXAMPLE: Table 1 shows the effects of light and hormones on the germination of light-sensitive lettuce seeds. One group of seeds was treated with ABA and left under a fluorescent light, while another group received GA in the dark. It can be seen that the majority of the seeds (92.5%) germinated in the light, but fewer germinated in the light when ABA was present.

The first sentence of this passage is unnecessary, because **it says nothing substantive about the data contained in Table 1**; instead, it describes the layout of the table, which needs no description. The second sentence is also unnecessary, because it restates the materials and methods, but says nothing about the results. The first part of the third sentence ("It can be seen that") is simply verbiage.

Avoid other unnecessary introductions such as:

- It was found that…
- When the test plates were observed and counted, …
- There is also a general trend where…
- It can be determined that…
- Table 1 shows that there is a relationship between hormones and seed germination. (*What* is the relationship?)

Scientific writing is concise and to the point. When you write the body of the Results section, eliminate unnecessary introductions. Instead, jump right in with a substantive observation. Write in past tense, because in scientific writing, present tense denotes that the statement is a generally accepted fact (see "Present Tense or Past Tense?" in Chapter 5). It remains to be seen if your results are accepted by the scientific community!

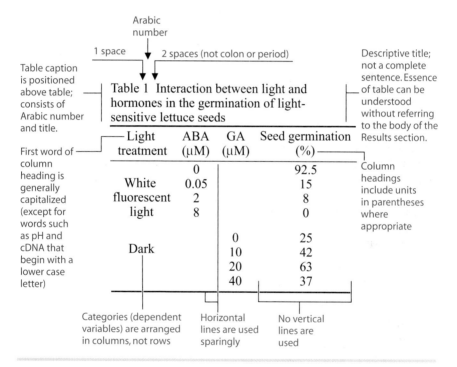

Text describing the table immediately precedes the table. Skip a line before and after the table.

Abscisic acid (ABA) and gibberellic acid (GA) had opposite effects on seed germination (Table 1). The majority of the seeds (92.5%) germinated in the light, but fewer germinated when they were exposed to 0.05-8 μM ABA. On the other hand, only 25% of the seeds germinated in the dark. This percentage increased when the seeds were exposed to 10-40 μM GA.

Refer to the table number parenthetically

Arabic number

1 space   2 spaces (not colon or period)

Table caption is positioned above table; consists of Arabic number and title.

Descriptive title; not a complete sentence. Essence of table can be understood without referring to the body of the Results section.

Table 1 Interaction between light and hormones in the germination of light-sensitive lettuce seeds

First word of column heading is generally capitalized (except for words such as pH and cDNA that begin with a lower case letter)

Column headings include units in parentheses where appropriate

Light treatment	ABA (μM)	GA (μM)	Seed germination (%)
White fluorescent light	0		92.5
	0.05		15
	2		8
	8		0
Dark		0	25
		10	42
		20	63
		40	37

Categories (dependent variables) are arranged in columns, not rows

Horizontal lines are used sparingly

No vertical lines are used

Figure 4.2   Excerpt from a Results section showing a properly formatted table preceded by the text that describes the data in the table

REVISION: The majority of the seeds (92.5%) germinated in the light, but fewer seeds germinated when they were exposed to 0.05-8 μM ABA (Table 1).

Visuals

The most common visuals in scientific writing are tables and figures. A
table is defined by Webster's dictionary as "a systematic arrangement of
data usually in rows and columns for ready reference." A **figure** is any
visual that is not a table. Thus, line graphs, bar graphs, pie graphs, drawings, gel photos, X-ray images, and microscope images are all called *figures* in scientific papers.

Do not feel that you have to have visuals in your lab report. If you can
state the results in a sentence, then **no visual** is needed (see "Organizing
Your Data," Example 1).

Tables. Tables are used to display large quantities of numbers and other
information that would be tedious to read in prose. Arrange the categories vertically, rather than horizontally, as this arrangement is easier
for the reader to follow (see, for example, Table 1 in Figure 4.2). List the
items in a logical order (e.g., sequential, alphabetical, or increasing or
decreasing value). Include the units in each column heading to save
yourself the trouble of writing the units after each number entry in the
table.

By convention, tables in scientific papers do not have vertical lines to
separate the columns, and horizontal lines are used only to separate the
table caption from the column headings, the headings from the data,
and the data from any footnotes.

Give each table a caption that includes a number and a title. Center
the caption or align it on the left margin *above* the table. Use Arabic
numbers, and number the tables consecutively in the order they are discussed in the text. Notice that in this book, the table and figure numbers
are preceded by the chapter number. This system helps orient the reader in long manuscripts, but is not necessary in short papers like your
laboratory report.

Titles consist of a precise noun phrase, not a complete sentence, and
only the first word and proper nouns are capitalized. The reader should
be able to understand the essence of the table without having to refer to
the body (the text) of the Results section.

FAULTY: Table 1 Table of interaction between light and
hormones in the germination of light-sensitive
lettuce seeds [Do not start a title for a visual
with a description of the visual.]

FAULTY: Table 1 Seed germination data [Do not write
vague and undescriptive titles.]

REVISION: Table 1 Interaction between light and hor-
             mones in the germination of light-sensitive let-
             tuce seeds

A table is always positioned *after* the text in which you refer to it (see
Figure 4.2). Mention the table number in parentheses at the end of the
first sentence in which you describe the table contents. That way the
reader can refer to the table as you describe what you consider to be
important.

In your laboratory report, it is not necessary to include a table when
you already have a graph that shows the same data. Make *either* a table
or a graph—not both—to present a given data set.

Tables can be constructed in either Microsoft Word (see Appendix 1,
Section 2.12) or Microsoft Excel (see "Importing Tables Into Microsoft
Word" in Appendix 2).

Line graphs. Line graphs (or XY graphs) are perhaps the most frequently used type of graph in biology. Line graphs are used to display a
trend or an important relationship between one or more variables. The
data displayed in line graphs are **continuous and numerical**.

By convention, the independent variable (the one the scientist manipulates) is plotted on the x-axis, and the dependent variable (the one that
changes in response to the independent variable) is plotted on the y-axis.
The data then show the effect of x on y, or y as a function of x.

There are a number of formats for plotting data on an XY graph
(Figure 4.3). Situations in which each format might be used are given.

Figures are always numbered and titled *beneath* the visual (see Figures
4.1 and 4.4). The captions may be centered or placed flush on the left
margin of the report. Arabic numbers are used, and the figures are numbered consecutively in the order they are discussed in the text.

Figure titles should consist of a precise noun phrase, not a complete
sentence, and only the first word and proper nouns are capitalized. The
reader should be able to understand the title without referring to the text
in the Results section.

FAULTY: Figure 1 Percentage of male gametophytes vs.
           population density [Do not restate the y-axis
           label versus the x-axis label as the figure title.]

FAULTY: Figure 1 shows the effect of population density
           on the percentage of male gametophytes in
           wild type *Ceratopteris* [Separate the figure
           number and the title.]

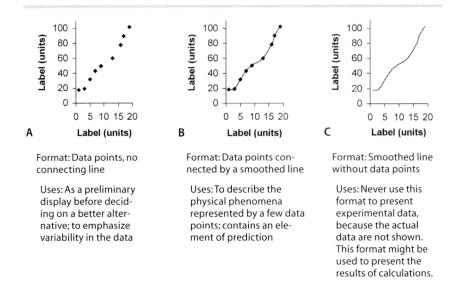

A   Format: Data points, no connecting line

Uses: As a preliminary display before deciding on a better alternative; to emphasize variability in the data

B   Format: Data points connected by a smoothed line

Uses: To describe the physical phenomena represented by a few data points; contains an element of prediction

C   Format: Smoothed line without data points

Uses: Never use this format to present experimental data, because the actual data are not shown. This format might be used to present the results of calculations.

Figure 4.3 Different formats for XY graphs

FAULTY: Figure 1 Line graph of the effect of population density on the percentage of male gametophytes in wild type *Ceratopteris* [Do not start a title for a visual with a description of the visual.]

FAULTY: Figure 1 Averaged class data for C-fern experiment [Do not write vague and undescriptive titles.]

REVISION: Figure 1 Effect of population density on the percentage of male gametophytes in wild type *Ceratopteris*

If there is more than one data set (line) on the figure, you have three options:

- Add a brief label (no border, no arrows) next to each line
- Use a different symbol for each line and label the symbols in a key (as in Figure 4.4). Place the key without a border within the axes of the graph. This is the easiest option if you are using Excel to plot your data.

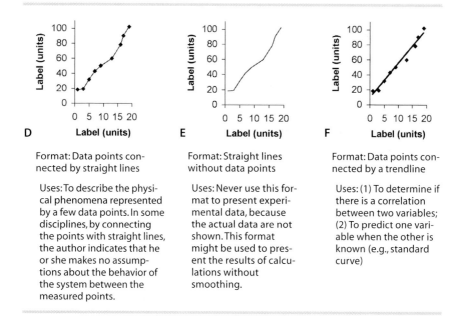

D Format: Data points connected by straight lines

Uses: To describe the physical phenomena represented by a few data points. In some disciplines, by connecting the points with straight lines, the author indicates that he or she makes no assumptions about the behavior of the system between the measured points.

E Format: Straight lines without data points

Uses: Never use this format to present experimental data, because the actual data are not shown. This format might be used to present the results of calculations without smoothing.

F Format: Data points connected by a trendline

Uses: (1) To determine if there is a correlation between two variables; (2) To predict one variable when the other is known (e.g., standard curve)

Figure 4.3 *Continued*

- If the first two options make the figure look cluttered, identify the symbols in the figure caption.

All three formats are acceptable in scientific papers as long as you use them consistently.

Figures in your laboratory report should be prepared according to the guidelines specified by the Council of Science Editors (CBE Manual, 1994). Although you may wish to plot a rough draft of your graphs by hand, you should learn how to use computer plotting software to make graphs. Microsoft Excel is a good plotting program for novices (see Appendix 2) because it is readily available and fairly easy to use. The time you invest now in learning to plot data on the computer will be invaluable in your upper-level courses and later in your career.

Bar graphs. A bar graph allows you to compare individual sets of data when these data are **non-numerical or discontinuous**—this is the main difference between line graphs and bar graphs. For example, if you wanted to compare the final height of the same species of plant treated with four different nutrient solutions (Figure 4.5), each bar could be used to represent the plants treated with one nutrient solution. The data

The percentage of male gametophytes increased
with population density of the wild type strain,
but there were no male gametophytes at any
density in the Her 1 strain (Figure 2).

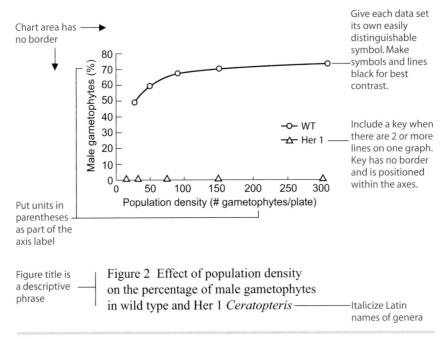

Chart area has → no border

Give each data set its own easily distinguishable symbol. Make symbols and lines black for best contrast.

Include a key when there are 2 or more lines on one graph. Key has no border and is positioned within the axes.

Put units in parentheses as part of the axis label

Figure title is a descriptive phrase

Figure 2 Effect of population density
on the percentage of male gametophytes
in wild type and Her 1 *Ceratopteris*

Italicize Latin names of genera

Figure 4.4 Excerpt from a Results section showing a properly formatted figure with two sets of data. A legend (key) is needed to distinguish the two lines. The text that describes the figure precedes it.

plotted on the *x*-axis are non-numerical, because different nutrient solutions (not numbers) are being compared. The data are discontinuous, because the individual solutions have no effect on the growth of the plants treated with the other nutrient solutions.

Bar graphs can be arranged either vertically (Figure 4.5, also called a column graph) or horizontally (Figure 4.6). The arrangement you use often depends on your sub-discipline, but the horizontal arrangement is more practical when the category labels are long.

The bars should be placed sequentially, but if there is no particular order, then put the control treatment bar far left in column graphs or at the top in horizontal bar graphs. Order the experimental treatment bars

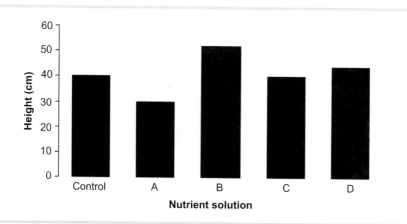

Figure 4.5 Final height of corn plants after 4-week treatment with different nutrient solutions. This figure is an example of a column graph.

from shortest to longest (or vice versa) to facilitate comparison among the different conditions. The baseline does not have to be visible, but all the bars must be aligned as if there were a baseline.

The bars should always be wider than the spaces between them. In a graph with clustered bars, make sure each bar has sufficient contrast so that it can be distinguished from its neighbor (Figure 4.7).

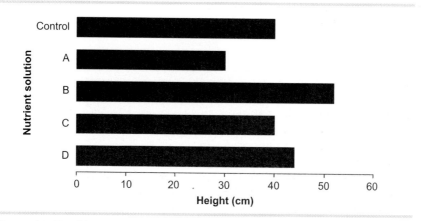

Figure 4.6 Final height of corn plants after 4-week treatment with different nutrient solutions. This figure is an example of a horizontal bar graph.

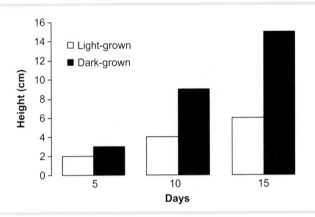

Figure 4.7 Difference in height of groups of light-grown and dark-grown bean seedlings at 5, 10, and 15 days after planting. This figure is an example of a clustered bar graph. Each bar in the cluster must be easily distinguishable from its neighbor.

Pie graphs. A pie graph is used to show data as a percentage of the total data. For example, if you were doing a survey of insects found in your backyard, a pie graph would be effective in showing the percentage of each kind of insect out of all the insects sampled (Figure 4.8). There should be between 2 and 8 segments in the pie. Place the largest segment in the right-hand quadrant with the segments decreasing in size clockwise. Combine small segments under the heading "Other." Position labels and percentages horizontally outside of the segments for easy reference.

Organizing Your Data

Reread the questions in the laboratory exercise to determine what your instructor expects you to learn from the data. You may also find specific instructions on how to organize the data (tables, line graphs, bar graphs, etc.). Based on these expectations, decide which visual is best. Ask yourself the following questions:

- Can I state the results in one sentence? If so, then **no visual** is needed.
- Are the numbers themselves more important than the trend shown by the numbers? If so, then use a **table**.
- Is the trend more important than the numbers themselves? If so, use a **graph**.

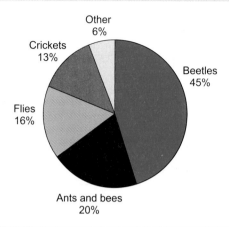

Figure 4.8 Composition of insects in backyard survey. Pie graphs are used to show data as a percentage of the total data.

The following examples demonstrate that there may be more than one good way to organize the data. Some visuals may be more appropriate than others, and in some cases, no visual may be the best alternative.

EXAMPLE 1: *Brassica* seeds were placed on filter paper saturated with pH 1, 2, 3, or 4 buffered solutions. The positive control was filter paper saturated with water. After 2 days, 100% of the seeds in the positive control germinated. No seeds germinated in any of the buffered solutions.

POSSIBLE
SOLUTION: **No visual** is needed because the results can be summarized in one sentence: "After 2 days, 100% of the seeds that imbibed water germinated, but none of the seeds that were treated with buffered solutions pH 1, 2, 3, or 4 germinated."

EXAMPLE 2: Light-sensitive lettuce seeds placed on filter paper saturated with water were exposed to the same fluence of white fluorescent, red, far-red, green, and blue light treatments as well as darkness, and the percentage that germinated was determined 30 hours later.

(A)

Table 1 Effect of light treatment on percentage of
light-sensitive lettuce seeds germinated after 30 hr

Light treatment	Seed germination (%)
Red	76
White	65
Blue	44
Far-red	38
Green	37
Dark	30

(B)

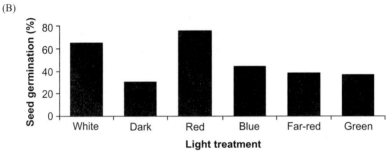

Figure 1 Effect of light treatment on percentage of light-sensitive lettuce
seeds germinated after 30 hr

Figure 4.9 Example 2 data summarized in (A) a table or (B) a bar graph. The
positive and negative controls are placed to the left, and, if there is no particular order to the categories (colors in this example), arrange the bars in order of
the longest to the shortest (or vice versa).

POSSIBLE
SOLUTION: Since some of the data are qualitative rather
than quantitative (colors rather than wavelengths of light), either a *table* or a *bar graph*
(Figure 4.9) works well to display the results.

INAPPROPRIATE: A *line graph* is **not** appropriate because the
exact wavelengths of light to which the seeds
were exposed are not known. *Text only* is **not**
appropriate because listing the seed germination percentages in a sentence is tedious to read
and hard to comprehend.

EXAMPLE 3:   Blackworms were exposed to different concentrations of a caffeine solution for 15 minutes.
The pulsation rate of the dorsal blood vessel
was measured before and after treatment and
the raw data were expressed as percent change
in pulsation rate (negative if the rate decreased
and positive if it increased).

POSSIBLE
SOLUTION: Whenever *different concentrations* of the same
drug, hormone, or solution are tested, think **dose-
response curve**. A dose-response curve is a *line
graph* that depicts the dose (independent variable)
on the *x*-axis and the response (dependent variable)
on the *y*-axis (Figure 4.10). The data points were
purposely not connected to show variability in
pulsation rate at the same concentration of caffeine
(due to variability among blackworms or variability
among student measurement of pulsation rate).

EXAMPLE 4:   The activity of an enzyme (catalase) was monitored
at nine different temperatures in order to determine
the optimal temperature for maximum activity.

POSSIBLE
SOLUTION: If your emphasis is on the actual numbers rather
than the trend, then display the results in a
table. If the trend is more important than the
numbers, use a *line graph and connect the points*
with smoothed or straight lines (Figure 4.11).

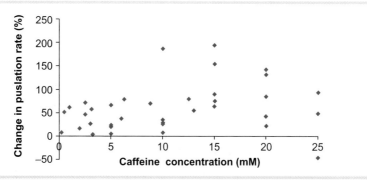

Figure 4.10 Effect of caffeine on pulsation rate of blackworms. When the data points are not connected in order to show variability, the line graph may also be called a scatter plot.

(A)

Table 1 Effect of temperature on catalase activity

Temperature (°C)	Catalase activity (units of product formed · sec^{-1})
4	0.039
15	0.073
23	0.077
30	0.096
37	0.082
50	0.04
60	0.007
70	0
100	0

(B)

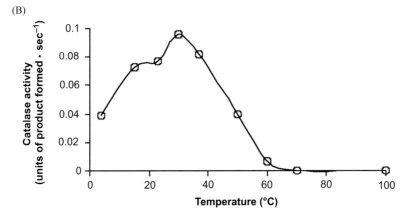

Figure 1 Effect of temperature on catalase activity

Figure 4.11 Example 4 data summarized in (A) a table or (B) a line graph. When data are summarized in a table, the emphasis is placed on the numbers rather than on the trend. For the same data, a line graph is more effective than a table in showing the trend.

Standard curves. Standard curves represent a *special case of line graph* in which the goal of the analysis is to predict one variable when the other is known. At first the data points are plotted as a scatter plot, and then a linear regression line (trendline) is fitted to the points. The degree to which the trendline actually fits the data is given by the R-squared value, whereby the closer the R-squared value is to 1, the better the correlation or fit.

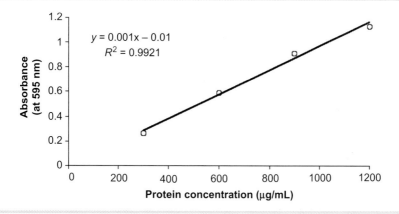

Figure 4.12 Bradford standard curve constructed with absorbance data for known concentrations of BSA. The R^2 value indicates that the linear regression line fits the data very well. The absorbance of a solution with unknown protein concentration is then measured, and the absorbance is substituted into the regression equation for y to determine x (protein concentration).

To make a standard curve for a protein assay, such as the Biuret assay or the Bradford assay, a scientist prepares known concentrations of a standard protein solution (such as bovine serum albumin) and then measures the absorbance of each solution using a spectrophotometer. The relationship between absorbance and concentration is expected to be linear (Figure 4.12), as expressed in Beer's law.

To predict the protein concentration in an "unknown" solution, the scientist measures the absorbance of the solution, and then predicts the protein concentration using the regression equation of the standard curve.

Think Ahead to the Discussion

As you describe the important findings of your investigation, think about what you already know about the subject. Are the results expected? Do they agree with the findings of other investigators?

- If **yes**, then you can jot down your ideas to use in the Discussion section.

- If no, try to develop possible explanations for the results. Some reasons to consider are:

  - Human error (failure to follow the procedure, failure to use the equipment properly, failure to prepare solutions correctly, variability when multiple lab partners measure the same thing, and simple arithmetic errors). If you suspect that human error may have influenced your results, it is important to acknowledge its contribution.

- Numerical values were entered incorrectly in the computer plotting program.
- Sample size was too small.
- Variability was too great to draw any conclusions.

If you can rule out these possibilities, discuss your results with your lab partner, teaching assistant, or instructor. If there is an obvious error, an "outsider" may be able to spot it immediately. Furthermore, informal discussions may help you clarify what you know and what you do not know about the topic.

Equations

Equations are neither tables nor figures. They should be set off from the rest of the text on a separate line. If you have several equations and need to refer to them unambiguously in the body of the Results (or other) section, number each equation sequentially and place the number in parentheses on the right margin. For example:

$$\text{Absorbance} = -\log T \tag{1}$$

If you are presenting a sequence of calculations, align the = symbol in each line, as in the following example.

Protein concentration of the unknown sample was determined using the equation of the Biuret standard curve. The measured absorbance value was substituted for y, and the equation was solved for x (the protein concentration):

$$y = 0.0417x$$
$$0.225 = 0.0417x$$
$$5.40 = x$$

Thus, the protein concentration of the sample was 5.40 mg/mL.

Make Connections

Now that the "meat" of your report is done, it's time to describe how your work fits into the existing body of knowledge. These connections are made in the Discussion and Introduction sections.

Write the Discussion

The Discussion section gives you the opportunity to **interpret your results and explain why they are important**. For a strong Discussion section:

- Summarize the results in a way that provides evidence for your conclusions. Avoid using the word "prove" when discussing your results:

    FAULTY: These results prove that my hypothesis is correct.

    REVISION: These results provide support for my hypothesis.

- State how your results relate to existing knowledge (cite literature sources).
- Point out any inconsistencies in your data. This is preferable to concealing an anomalous result.
- Discuss possible sources of error.
- Describe future extensions of the current work.

Write the Introduction

The Introduction concisely states what motivated the study, how it fits into the existing body of knowledge, and the objectives of the work. The introduction consists of two primary parts: (1) background information from the literature and (2) objectives of the current work.

After having written drafts of the Materials and Methods, Results, and Discussion sections, you should be intimately familiar with the procedure, the data, and what the data mean. Now you are in a position to put your investigation into perspective. What was already known about the topic? Were there any inconsistencies or unanswered questions? Why did you carry out this investigation?

If you designed your own experiment based on the scientific method, you probably answered these questions already. If this investigation was a laboratory exercise prepared by your instructor, reread the exercise to see if some of these questions have already been answered. The idea is not to copy the laboratory exercise introduction verbatim, but to make sure you understand the objectives.

Use keywords from the objectives to locate background information in the literature (see Chapter 2). Evaluate the titles in the "hits" critically to make sure the sources are relevant to your investigation.

The opening sentence of the Introduction section is usually a general observation or result familiar to the audience. By convention, generally accepted statements are written in present tense (see "Present Tense or

Past Tense?" in Chapter 5). Subsequent sentences narrow down the topic to the specific focus of the current investigation. Subsequent paragraphs then provide background information from the literature and describe unanswered questions or inconsistencies. The objectives of the current work are usually stated in past tense in the last paragraph of the Introduction section.

Effective Advertising

The whole point of writing your paper is to communicate your work to your peers. The Abstract and the Title are the primary tools your audience will use to decide whether or not they are interested in your work.

Write the Abstract

The abstract is a **summary of the entire paper** in 250 words or less. It contains:

- An introduction (scope and purpose)
- A short description of the methods
- The results
- Your conclusions

There are no literature citations or references to figures in the Abstract.

After the title, the Abstract is the most important part of the scientific paper used by the reader to determine initial interest in the author's work. Abstracts are indexed in databases that catalogue the literature in the biological sciences. If an abstract suggests that the author's work may be relevant to your own work, you will probably want to read the whole article. On the other hand, if an abstract is vague or essential information is missing, you will probably decide that the paper is not worth reading. When you write the Abstract for your own laboratory report, put yourself in the position of the reader. If you want the reader to be interested in your work, write an effective Abstract.

Writing the Abstract is difficult because you have to condense your entire paper into 250 words or less. One strategy for doing this is to list the key points of each section, as though you were taking notes on your own paper. Then write the key points in full sentences. Revise the draft for clarity and conciseness using strategies such as using active voice, combining choppy sentences with connecting words, rewording run-on sentences, and eliminating redundancy. With each revision, look for ways to shorten the text so that the resulting Abstract is a concise and accurate summary of your work.

The ability to write abstracts is important to a scientist's career. Should you someday wish to present your research at an academic society meeting, such as the Society for Neuroscience, the American Association for the Advancement of Science, or the National Association of Biology Teachers (to name just a few), you will be asked to submit an abstract of your presentation to the committee in charge of the meeting program. Your chances of being among the select field of presenters at these meetings are much better if you have learned to write a clear and intelligent abstract.

Write the Title

The Title is a **short, informative description** of the essence of the paper. You may choose a working title when you begin to write your paper, but revise the title after subsequent drafts. Remember that readers use the title to determine initial interest in the paper, so descriptive accuracy is the most essential element of your title. Brevity is nice if it can be achieved. Some journals (especially the British ones) are fond of puns and humor in their titles, but this kind of thing may be better left for later in your career.

Here are some examples of vague and nondescriptive titles:

> FAULTY: Quantitative Protein Analysis

> FAULTY: The Assessment of Protein Content in an
> Unknown Sample

> FAULTY: Egg White Protein Analysis

These titles leave the reader wondering what method of protein analysis was used and what sample was analyzed.

> REVISION: Assessment of Protein Concentration in Egg
> White Using the Biuret Method

Documenting Sources

References are typically cited in the Introduction and Discussion sections of a scientific paper, and the procedures given in Materials and Methods are often modifications of those in previous work. Citation and reference format in the sciences differs from that in the humanities in two important ways:

- It is not customary to use direct quotations in scientific papers; paraphrase instead.

- In the citation in the text, it is not customary to give the page number(s) of the source; instead, use one of the two citation systems explained in the following section.

The References section is also called *Works Cited* or *Literature Cited*, because it includes only the published sources you cited in your paper. It is not a "bibliography" or a list of works consulted to learn more about the topic.

Citation and Reference Formats

Whenever you use another person's ideas, whether they are published or not, you must document the source. This is done by citing the source in an abbreviated form in the text and then giving the full reference in the References section at the end of the document. An exception to this practice is personal communications, which are cited in the text, but are not listed under References. Only sources that have been cited in the text may be included in the References section.

The CBE Manual (1994) recommends using the Citation-Sequence System (C-S) or the Name-Year System (N-Y) for documenting your sources. The system you actually use depends on your instructor's preference or on the format specified by the particular scientific journal in which you aspire to publish.

Examples of the same passage of text documented using the C-S and N-Y formats are given in Table 4.3. Notice in both examples how inconspicuously the source is documented. This style is quite different from that practiced in some disciplines in the humanities, in which the source may be introduced at length in the text—for example: "According to Warne and Hickock in their 1989 paper published in *Plant Physiology*, antheridiogen may be related to the gibberellins." **Do not use this style in your lab reports!**

With regard to the full reference, the two formats differ in the sequence of information and the listing of the month of publication. In the N-Y system, the year of publication follows the author's name; in the C-S system, the year follows the journal name. The month of publication is only used in C-S format.

The Name-Year system has the advantage that people working in the field will know the literature and, on seeing the authors' names, will understand the reference without having to check the reference list. With the citation-sequence system, the reader must turn to the reference list at the end of the paper to gain the same information. The important thing, however, is to choose a citation format and adhere to it consistently throughout your paper.

The Citation-Sequence System

In the Citation-Sequence (C-S) system (Table 4.3, p. 60), the source is cited with a number in parentheses following the paraphrased sentence (in some journals, this number is in square brackets or displayed as a superscripted endnote, but never a footnote). On the page of references at the end of the document, the citations are listed in **numerical order** and include the full reference.

The Name-Year System

In the Name-Year (N-Y) system (Table 4.3, p. 61), the source is given in the form of author(s) and year. The source may be cited in parentheses or as the subject of the sentence, as shown in the following examples:

PARENTHESES: C-fern gametophytes respond to antheridio-gen only for a short time after inoculation (Banks and others 1993).

AS THE SUBJECT: Banks and others (1993) found that C-fern gametophytes respond to antheridiogen only for a short time after inoculation.

The **number of authors** determines how the authors are cited in N-Y system (Table 4.4).

On the page of references at the end of the document, the references are listed in **alphabetical order** according to the first author's last name. The last name is written first, followed by the initials. When there are 10 or fewer authors, list all authors' names. When there are more than 10 authors, list the first 10 and then write *et al.* or *and others* after the tenth name.

The reference styles for the Citation-Sequence and Name-Year formats are summarized in the boxes on pages 62 and 63.

Electronic Journal Articles

The citation in the text is the same as for printed sources:

C-S   ...confirmed in a previous study (1).

N-Y   ...was confirmed in a previous study by Banks (1994).

The full reference at the end of the document begins with the same information that would be provided for a printed source. The Web information

TABLE 4.3 Two systems of source documentation used in scientific papers

CITATION-SEQUENCE SYSTEM	
	TEXT CITATIONS
Sources are listed in the order they are cited	Gametophytes of the tropical fern *Ceratopteris richardii* (C-fern) develop either as males or hermaphrodites. Their fate is determined by the pheromone antheridiogen (1, 2). Gametophytes respond to antheridiogen only for a short time between 3 and 4 days after inoculation (3). Although the structure of antheridiogen is unknown, it is thought to be related to the gibberellins (4). Gibberellins are a group of plant hormones involved in stem elongation, seed germination, flowering, and fruit development (5).
	CORRESPONDING FULL REFERENCES
Article in book	1. Näf U. Antheridiogens and antheridial development. In: Dyer AF, editor. The Experimental Biology of Ferns. London: Academic Press; 1979. pp. 436-470.
Journal article	2. Näf U, Nakanishi K, Endo M. On the physiology and chemistry of fern antheridiogens. Bot. Rev. 1975 Jul-Sep; 41(3): 315-359.
Journal article	3. Banks J, Webb M, Hickok L. Programming of sexual phenotype in the homosporous fern *Ceratopteris richardii*. Inter. J. Plant Sci. 1993 Dec; 154(4): 522-534.
Journal article	4. Warne T, Hickok L. Evidence for a gibberellin biosynthetic origin of *Ceratopteris* antheridiogen. Plant Physiol. 1989 Feb; 89(2): 535-538.
Book	5. Treshow M. Environment and Plant Response. New York: McGraw-Hill; 1970. 250 p.

Book references give the total number of pages in the book, not the pages from which you extracted the information

Reference Formats

TABLE 4.3 *Continued*

NAME-YEAR SYSTEM	
	TEXT CITATIONS
1 author 3 or more authors 2 authors 1 author	Gametophytes of the tropical fern *Ceratopteris richardii* (C-fern) develop either as males or hermaphrodites. Their fate is determined by the pheromone antheridiogen (Näf 1979; Näf and others 1975). Banks and others (1993) found that gametophytes respond to antheridiogen only for a short time between 3 and 4 days after inoculation. Although the structure of antheridiogen is unknown, it is thought to be related to the gibberellins (Warne and Hickok 1989). Gibberellins are a group of plant hormones involved in stem elongation, seed germination, flowering, and fruit development (Treshow 1970).
	CORRESPONDING FULL REFERENCES
Journal article	Banks J, Webb M, Hickok L. 1993. Programming of sexual phenotype in the homosporous fern *Ceratopteris richardii*. Inter. J. Plant Sci. 154(4): 522-534.
Article in book	Näf U. 1979. Antheridiogens and antheridial development. In: Dyer AF, editor. The Experimental Biology of Ferns. London: Academic Press. pp. 436-470.
Journal article	Näf U, Nakanishi K, Endo M. 1975. On the physiology and chemistry of fern antheridiogens. Bot. Rev. 41(3): 315-359.
Book	Treshow M. 1970. Environment and Plant Response. New York: McGraw-Hill. 250 p.
Journal article	Warne T, Hickok L. 1989. Evidence for a gibberellin biosynthetic origin of *Ceratopteris* antheridiogen. Plant Physiol. 89(2): 535-538.

Reference Formats

	C-S REFERENCE FORMAT
	The references are listed **in the order they are cited**. The author's last name is written first, followed by the initials. When there are 10 or fewer authors, list all authors' names. When there are more than 10 authors, list the first 10 and then write *et al.* or *and others* after the tenth name.
Journal article	Number of the citation. First author's last name First initials, Subsequent authors' names separated by commas. Article title. Journal title Year Month; Volume number(issue number): inclusive pages.
Article in book	Number of the citation. First author's last name First initials, Subsequent authors' names separated by commas. Article title. In: Editors' names followed by a comma and the word *editors*. Book title, edition. Place of Publication: Publisher; Year of publication. pp inclusive pages.
Book	Number of the citation. First author's or editor's last name First initials, Subsequent authors' or editors' names separated by commas. Title of book. Place of Publication: Publisher; Year of publication. Total number of pages in book followed by *p*.

is then placed at the end of the reference. It is absolutely essential to include the date accessed, because documents on the Web may move or be removed altogether.

C-S   1. Banks JA. Sex-determining genes in the homosporous fern *Ceratopteris*. Development 1994 Jul; 120(7): 1949-1958. <http://dev.biologists.org/cgi/ reprint/120/7/1949.pdf> Accessed 2003 May 24.

N-Y   Banks JA. 1994. Sex-determining genes in the homosporous fern *Ceratopteris*. Development 120(7): 1949-1958. <http://dev.biologists.org/cgi/reprint/120/7/1949.pdf> Accessed 2003 May 24.

N-Y REFERENCE FORMAT	

The references are listed in **alphabetical order**. The last name is written first, followed by the initials. When there are 10 or fewer authors, list all authors' names. When there are more than 10 authors, list the first 10 and then write *et al.* or *and others* after the tenth name. Type references with hanging indent format.

Journal article	First author's last name First initials, Subsequent authors' names separated by commas. Year of publication. Article title. Journal title Volume number(issue number): inclusive pages.
Article in book	First author's last name First initials, Subsequent authors' names separated by commas. Year of publication. Article title. In: Editors' names followed by a comma and the word *editors*. Book title, edition. Place of Publication: Publisher. pp inclusive pages.
Book	First author's or editor's last name First initials, Subsequent authors' or editors' names separated by commas. Year of publication. Title of book. Place of Publication: Publisher. Total number of pages in book followed by *p*.

Reference Formats

Sources from the Internet and the World Wide Web

Ease of access to the latest information makes the World Wide Web an irresistible source for nearly everything. Scientific communications are no exception. Take heed, however; information retrieved from the Internet and the World Wide Web may be unreliable. The authors of the communication, not journal editors or copy editors, are responsible for the accuracy of the contents. Because there is no review from an outside source, mistakes can occur. For this reason, it is preferable to cite only reliably published books or articles in your laboratory report.

In some instances, sources from the Web may be used to support published findings. Use one of the two citation formats described previously. If the author is not known, simply abbreviate the title of the Web page, and include the date of publication. If there is no date of publica-

TABLE 4.4 Number of authors determines how the authors are cited in N-Y system

NUMBER OF AUTHORS	AUTHOR AS SUBJECT	PARENTHETICAL REFERENCE (THE COMMA BETWEEN AUTHOR(S) AND YEAR IS OPTIONAL.)
1	Author's last name (year) found that...	(Author's last name, year)
2	First author's last name and second author's last name (year) found that ...	(First author's last name and second author's last name, year)
3 or more	First author's last name followed by and others or et al. (year) found that ...	(First author's last name and others, year) or et al. instead of "and others"

Note: If you cite more than one paper published by the same author in *different* years, list them in chronological order: (Dawson 2001, 2003). If you cite more than one paper published by the same author in the *same* year, add a letter after the year: "...was described in recent work by Dawson (1999a, 1999b)."

tion, cite the date of modification; if neither of these dates is given, use the date accessed. For example:

> C-S   ...is recommended by the Counselling Services at the University of Victoria (1).

> N-Y   ...is recommended by the Counselling Services at the University of Victoria (*How to Read University Texts*, 2001).

The full reference at the end of the document would follow the format given in "Electronic Journal Articles," with the same information that would be provided for a printed source plus the URL and the date accessed.

When URLs are used in text, they are enclosed in angle brackets (< >) to distinguish them from the rest of the text. (URLs can also be printed in color, in which case the angle brackets are not required.) Every character in a URL is significant, as are spaces and capitalization. Very long URLs can be broken before a punctuation mark (tilde ~, hyphen -, underscore _, period ., forward slash /, backslash \, or pipe |). The punctuation mark is then moved to the next line.

For more information on using Internet sources, see Harnack and Kleppinger (2001).

Unpublished Laboratory Exercise

Unpublished material is usually not included in the References section. However, if your instructor asks that you cite laboratory exercises in your laboratory report, the format could look like this:

> C-S   Author ([*Anonymous*] *if unknown*). Title of lab exercise. Course number, Department, University. Year.

> N-Y   Author ([*Anonymous*] *if unknown*). Year. Title of lab exercise. Course number, Department, University.

Personal Communication

If you had a discussion with your professor or a colleague, and you obtained certain information from him or her that you could not find in a published source, it is appropriate to recognize this person's contribution as follows:

> …may be explained by possible contamination from a virus or bacterium (M. Pizzorno, personal communication, 6 Nov 2003).

It is **not necessary** to include personal communications in the References section.

Reference Formats

Chapter 5

REVISION

Revision—reading your paper and making corrections and improvements—is an important task that usually does not get nearly the attention it deserves. Too many students write the first draft of their laboratory report the night before it is due and hand in the hard copy, still warm from the printer, without even having proofread it.

The truth is, most writers cannot produce a clear, concise, and error-free product on the first try. It may take several revisions before the writer is satisfied that he or she has conveyed, with clarity and logic, the motivation for writing the paper, the important findings, and the conclusions. For this reason, **prepare the first draft as early as possible** so that you will have the time to think about it, get help if necessary, get feedback from your peer reviewer, and make final revisions.

Getting Ready to Revise

Take a Break

The first step in revision is *not* to do it immediately after you have written the first draft. You need to distance yourself from the paper to gain the objectivity needed to read the paper critically. So take a break, and get a good night's sleep.

Look at the Big Picture

In the early stages of the revision process, content is more important than format. Make sure you have answered all the questions in the laboratory exercise (if applicable), and use the "Biology Lab Report Checklist" (pp. 96–97) to make sure each section of your laboratory report contains the proper content. Don't worry about whether or not a table caption is in the right position until you are sure you even need to include the table.

Ask yourself the following questions:

- What are the goals and objectives of this paper? Did I make these objectives clear to the reader?
- What questions and concerns do I have with this draft? What are some ways I can begin to address these concerns?
- What parts need feedback (not sure if my meaning is clear, not sure if I understood this concept correctly, and other issues)?
- What do I like about this paper? What are its strengths and what has gone well?

Be prepared to discuss your answers to these questions with your peer reviewer. Peer review is the next step in the revision process.

Get Feedback

When we are engrossed in our work, we may fail to recognize that what is obvious to us is not obvious to an "outsider." Some typical problem areas are flow and organization of the paper and typographical or grammatical errors.

That is where feedback from someone who is familiar with the subject matter comes in handy. Ask your lab partner or another classmate to review your paper. Return the favor by reviewing his or hers. The questions your reviewer will focus on are as follows:

- Do I know what the writer is trying to accomplish with this paper? Is the purpose clear?
- What questions or concerns do I have about this paper? Are there sections that were difficult to follow? Are the organization, content, flow, and level appropriate for the intended audience?
- What suggestions can I offer the writer to help him/her clarify the intended meaning?
- What do I like about the paper? What are its strengths?

Tips for being a good peer reviewer. There are two issues with which you may struggle when you are asked to review your classmate's paper: (1) I'm not confident that I know the "right" answer or know enough about the writing process to give good suggestions, and (2) I don't want to hurt the writer's feelings. These are valid concerns, and resolving them will require, first, a willingness to learn as much about writing scientific papers as possible, and second, the attitude that if something is unclear to you, it may also be unclear to other readers. With each paper you review, you will gain more confidence in your ability to give constructive feedback. In the meantime, however, a good rule of thumb is to give the kinds of suggestions and consideration that you would like to receive on your own paper.

There are two schools of thought concerning how the reviewer should provide feedback on the paper. One is that no marks should be made on the paper itself; instead, any comments should be written on a separate sheet of paper. The second school of thought is just the opposite. Although the first method respects the writer's ownership of the paper, it is not very practical. First of all, it is difficult to describe exactly where on the paper the revision should be made. Secondly, vague suggestions like "improve flow of introduction" are not nearly as helpful to your classmate as "what do you mean by..." and then circle or underline the confusing phrase. Think of the peer review process as a team sport: you are not challenging your teammate's right to be on the team. You are working together to get the best possible result.

If it is not possible for you and your classmate to exchange lab reports in person, you might consider sending your paper by e-mail as an attached file. To see the changes made by your peer reviewer, use the Track Changes feature in Word (See "Tracking Changes made by Peer Reviewers" in Appendix 1).

Here are some concrete tips for being a good peer reviewer:

- Talk to the writer about his or her objectives, questions and concerns, parts that need specific feedback, and perceived strengths and weaknesses.
- Use the "Biology Lab Report Checklist" for content.
- Use a pencil and standard proofreader's marks to mark awkward sentences, spelling and punctuation mistakes, and formatting errors. Do not feel you have to rewrite individual sentences—that is the writer's job.
- Ask questions. Let the writer know where you can't follow his/her thinking, where you need more examples, where you expect more detailed analysis, and so on.
- Do not be embarrassed about making lots of comments; the author does not have to accept your suggestions. On the other hand, if you say only good things about the paper, how will the writer know whether the paper is accomplishing the desired objectives?

You can fine-tune your proofreading skills on any text. You may recognize some of your own problems in other people's writing, and with persistence and practice, you will find creative solutions to correct these problems. Keep a log of the problems that recur in your writing and review them from time to time. Repetition builds awareness, which will help you achieve greater clarity in your writing.

Have an informal discussion with your peer reviewer. Sometimes the comments made by the peer reviewer on the paper are self-explanatory. Other times, however, the peer reviewer cannot respond to certain parts of the paper, because more information is required. Under these circumstances, an informal discussion between the writer and the reviewer is helpful. There are two important rules for this discussion:

- First, the writer talks and the reviewer listens. The objective is to help the writer express exactly what he/she wants to say in the paper.

- Second, the reviewer talks, in nonjudgmental terms, about which parts of the paper were readily understandable and which parts were confusing. The reviewer does not have to be an experienced writer to do this—no two people have exactly the same life experiences, and there is always something positive you can learn from looking at your writing from someone else's perspective.

Revise Again

The comments provided by your peer reviewer, both on your paper and during the subsequent informal discussion, should give you some ideas on how your paper can be improved. The areas that require attention will most likely fall into one of the following categories:

- Conventions in biology
- Numbers
- Standard abbreviations
- Punctuation
- Clarity
- Grammar
- Word usage
- Spelling
- Global revision

Conventions in Biology

Audience. Scientific papers are written by scientists for fellow scientists. When you write your laboratory reports, write for your fellow students as though they are scientists, not students in a classroom setting. Even though your instructor will be the one who reads and evaluates your reports, think of him or her as a mentor, not the audience.

> EXAMPLE: The purpose of this experiment ~~is to teach the~~ ~~student how~~ was to determine the protein concentration in an unknown sample.

The audience has a knowledge base similar to the author of the paper. Thus, when you introduce your topic and describe and interpret your results, you should assume that your audience will know some scientific vocabulary, but that you should clarify or define less-familiar terms. When deciding on how much background information to include, assume that your audience knows what you learned in class.

Do not use jargon (terms known only to experts) or copy sections of journal articles because you think that your own words are inferior to those of experts in the field. The objective of your writing is not to impress your instructor with empty words, but to enable your peers to understand what you have accomplished.

Introductions. In some fields, the author attempts to attract a large audience for his or her paper by asking an intriguing question or presenting a dilemma to capture the reader's interest. In scientific papers, however, the author assumes that the audience is already interested in the subject. Readers of scientific papers typically determine initial interest in a paper from the title and the abstract. If these sections seem promising, the reader will then peruse the Introduction to determine further interest. There is no need for the author to pique the reader's interest with longwinded introductions, because the interest is already there.

When you write the Introduction section of your laboratory report, just jump right in with your topic. Eliminate unnecessary introductions such as the first sentence in the following example:

> EXAMPLE: ~~In many areas of biology, it is important to gain a~~ ~~higher understanding of enzyme activity.~~ Rate of enzyme activity depends on factors such as enzyme concentration, substrate concentration, pH, temperature, and the presence of inhibitors.

Get to the point. Describe what area of enzyme activity you studied, why, and what you expected to learn from your work.

You may be tempted to go overboard with detail in the Results section of your report, as illustrated by the following example:

> EXAMPLE: ~~From the data that has been gathered, a graph~~
> ~~depicting the effect of various pH environments~~
> ~~on the rate of catalase activity is represented.~~
> ~~The graph displays the pH tested and the reac-~~
> ~~tion rate. The data plotted is an accumulation of~~
> ~~data from several lab sections. Through analyz-~~
> ~~ing the graph I can see that~~ ~~T~~there ~~is~~ <u>was</u> no
> activity below a pH of 4 or above 10. Maximum
> catalase activity occurred at pH 7.

Cut to the chase—don't waste the reader's time with sentences that say nothing substantive.

Initially it is difficult to write in (and read) the terse, get-to-the-point style that characterizes scientific papers. With practice, however, you may come to appreciate this style, because in a well-written paper, not a word is wasted. The benefit to you as a reader is that you extract a maximum amount of information from a minimum amount of text.

Present tense or past tense?   In scientific papers, present tense is used mainly in the following situations:

- To make generally accepted statements (for example, "Photosynthesis *is* the process whereby green plants produce sugars.")
- When referring directly to a table or figure in your paper (for example, "Figure 1 *is* a schematic diagram of the apparatus.")
- When you state the findings of published authors (for example, "Catalase HPII from *E. coli is* highly resistant to denaturation (Switala and others, 1999).")

Past tense is used mainly in the following situations:

- To report your own work, especially in the Abstract, Materials and Methods, and Results sections, because it remains to be seen if it is accepted as fact (for example, "At temperatures above 37°C, catalase activity *decreased*.").
- To cite another author's findings directly (for example, "Miller and others (1998) *found* that…").

Active and passive voice.   In **active voice**, the subject *performs* the action. In **passive voice**, the subject *receives* the action. Consider the following example:

PASSIVE: The clam was opened by the sea star.
               [Emphasis on *clam*]

ACTIVE: The sea star opened the clam.
               [Emphasis on *sea star*]

Although the meaning is the same in both sentences, notice the difference in emphasis. In active voice, the emphasis is on the performer, and the action takes place in the direction the reader reads the sentence. Active voice is recommended by most style guides for reasons that include the following:

- It sounds more natural and is easier for the reader to process.
- It is shorter and more dynamic.
- There is no ambiguity about who/what the subject of the sentence is, or about who did the action.

Consider the following example:

PASSIVE: It was concluded from this observation that…

ACTIVE: I concluded from this observation that…

Passive voice leaves the reader wondering who drew the conclusion; active voice conveys this information clearly.

While active voice is generally preferred, passive voice may be more appropriate when *what* is being done is more important than *who* is doing it. For example:

PASSIVE: Catalase was extracted from a potato.
               [Emphasis on *catalase*]

ACTIVE: I extracted catalase from a potato.
               [Emphasis on *I*]

Notice the difference in emphasis. Is it really important to the success of the procedure that *you* did it, or does the emphasis belong on the catalase?

A paper that contains a mixture of active and passive voice is pleasant to read. Your decision to use active or passive voice in a sentence should ultimately be determined by clarity and brevity. In other words, use active voice to emphasize the subject and the fact that the subject is performing the action. Use passive voice when the action is more important than who is doing it.

Numbers

Numbers are used for quantitative measurements. In the past, numbers less than 10 were spelled out, and larger numbers were written as numerals. The modern scientific number style recommended in the upcoming Seventh Edition of the CBE Manual (<http://www.councilscienceeditors.org/publications/ssf_numberstyle.cfm>, accessed 2004 Jan 31) aims for a more consistent usage of numbers. The new rules are as follows:

1. Use numerals to express any *quantity*. This form increases their visibility in scientific writing, and emphasizes their importance.
   - Cardinal numbers, for example, 3 observations, 5 samples, 2 times
   - In conjunction with a unit, for example, 5 g, 0.5 mm, 37°C, 50%, 1 hr. Pay attention to spacing, capitalization, and punctuation (see Table 5.1).
   - Mathematical relationships, for example, 1:5 dilution, 1000× magnification, 10-fold

2. Spell out numbers in the following cases:
   - When the number begins a sentence, for example, *"Fifty g of potatoes was [not were] homogenized."* rather than *"50 g of potatoes was homogenized."* Alternatively, restructure the sentence so that the number does not begin the sentence. Notice that when numbers are used in conjunction with units, the quantity is considered to be singular, not plural.
   - When there are two adjacent numbers, retain the numeral that goes with the unit, and spell out the other one. An example of this is *The solution was divided into four 250-mL flasks.*
   - When the number is used in a nonquantitative sense, for example, *one of the treatments, the expression approaches zero, one is required to consider…*
   - When the number is an ordinal number less than 10, and when the number expresses rank rather than quantity, for example, *the second time, was first discovered.*
   - When the number is a fraction used in running text, for example, *one-half of the homogenate, nearly three-quarters of the plants.* When the precise value of a fraction is required, however, use the decimal form, for example, *0.5 L* rather than *one-half liter.*

3. Use scientific notation for very large or very small numbers. For the number 5,000,000, write 5×10^6, not *5 million*. For the number *0.000005*, write 5×10^{-6}.

4. For decimal numbers less than one, always mark the ones column with a zero. For example, write *0.05*, not *.05*.

Standard Abbreviations

The CBE Manual (1994) defines standard abbreviations for authors and publishers in the sciences and mathematics. Some of the terms and abbreviations that you may encounter in introductory biology courses are shown in Table 5.1. Take note of spacing, case (capital or lowercase letters), and punctuation use. Except where noted, the symbols are the same for singular and plural terms (for example, 30 min *not* 30 mins).

Punctuation

The purpose of punctuation marks is to divide sentences and parts of sentences to make the meaning clear. A few of the most common punctuation marks and their uses are described in the following section. For a more comprehensive, but still concise, treatment of punctuation, see Hacker (1997), Lunsford (2001), or Lunsford and Connors (1995, 1999).

The comma. The comma inserts a pause in the sentence in order to avoid confusion. Note the ambiguity in the following sentence:

> While the sample was incubating the students prepared the solutions for the experiment.

A comma *should* be used in the following situations:

1. To connect two independent clauses that are joined by *and, but, or, nor, for, so,* or *yet.* An independent clause contains a subject and a verb, and can stand alone as a sentence.

   EXAMPLE: Feel free to call me at home, but don't call after 9 P.M.

2. After an introductory clause, to separate the clause from the main body of the sentence.

   EXAMPLE: Although she spent many hours writing her lab report, she earned a low grade because she forgot to answer the questions in the laboratory exercise.

   A comma is not needed if the clause is short.

   EXAMPLE: Suddenly the power went out.

3. Between items in a series, including the last two.

   EXAMPLE: Enzyme activity is affected by factors such as substrate concentration, pH, temperature, and salt.

TABLE 5.1 Standard abbreviations in scientific writing

TERM	SYMBOL OR ABBREVIATION	EXAMPLE
Latin words and phrases [The Council of Biology Editors Manual recommends that Latin words be replaced with English equivalents.]		[The Latin word may be replaced with the English equivalent given in brackets.]
circa (approximately)	ca.	The lake is ca. [approx.] 300 m deep.
et alii (and others)	*et al.*	Jones *et al.* [and others] (1999) found that …
et cetera (and so forth)	etc.	pH, alkalinity, etc. [and other characteristics] were measured.
exempli gratia (for example)	e.g.	Water quality characteristics (e.g., [for example,] pH, alkalinity) were measured.
id est (that is)	i.e.	The enzyme was denatured at high temperatures, i.e., the enzyme activity was zero. [Because the enzyme was denatured at high temperatures, the enzyme activity was zero.]
nota bene (take notice)	NB	NB [Important!]: Never add water to acid when making a solution.
Length		
nanometer (10^{-9} meter)	nm	*Note:* There is a space between the number and the abbreviation. There is no period after the abbreviation.
micron (10^{-6} meter)	µm	
millimeter (10^{-3} meter)	mm	
centimeter (10^{-2} meter)	cm	
meter	m	450 nm, 10 µm, 2.5 cm
Mass		
nanogram (10^{-9} gram)	ng	*Note:* There is a space between the number and the abbreviation. There is no period after the abbreviation.
microgram (10^{-6} gram)	µg	
milligram (10^{-3} gram)	mg	
gram	g	
kilogram (10^{3} gram)	kg	450 ng, 100 µg, 2.5 g, 10 kg

TABLE 5.1 *Continued*

TERM	SYMBOL OR ABBREVIATION	EXAMPLE
Volume		
microliter (10^{-6} liter)	μl or μL	*Note:* There is a space between the number and the abbreviation. There is no period after the abbreviation.
milliliter (10^{-3} liter)	ml or mL	
liter	l or L	
cubic centimeter (ca. 1 mL)	cm^3	450 μl or 450 μL, 0.45 ml or 0.45 mL, 2 l or 2 L
Time		
seconds	s or sec	*Note:* There is a space between the number and the abbreviation. There is no period after the abbreviation (unless the unit ends a sentence).
minutes	min	
hours	h or hr	
days	d	
		60 s or 60 sec, 60 min, 24 h or 24 hr, 1 d
Concentration		
molar (U.S. use)	M	TBS contains 0.01 M Tris-HCl, pH 7.4 and 0.15 M NaCl.
molar (SI units)	mol L^{-1}	
parts per thousand	ppt	Brine shrimp can be raised in 35 ppt seawater.
Other		
degree(s) Celsius	°C	15°C (no space between number and symbol)
degree(s) Fahrenheit	°F	59°F (no space between number and symbol)
diameter	diam.	pipe diam. was 10 cm
figure, figures	Fig., Figs.	As shown by Fig. 1, …
foot-candle	fc or ft-c	500 fc or 500 ft-c
maximum	max	The max enzyme activity was found at 36°C.
minimum	min	The min temperature of hatching was 12°C.
mole	mol	
percent	%	95% (no space between number and symbol)
species (sing.)	sp.	*Tetrahymena* sp.
species (plur.)	spp.	*Tetrahymena* spp.

4. Between coordinate adjectives (if the adjectives can be connected with *and*)

EXAMPLE: The students' original, humorous remarks made my class today particularly enjoyable. [*Original and humorous remarks* makes sense.]

A comma is not needed if the adjectives are cumulative (if the adjectives cannot be connected with *and*).

EXAMPLE: The three tall muscular students look like football players. [It would sound strange to say *three and tall and muscular students*.]

5. With *which,* but not *that* (see Word usage: that, which)

6. To set off conjunctive adverbs such as *however, therefore, moreover, consequently, instead, likewise, nevertheless, similarly, subsequently, accordingly,* and *finally*

EXAMPLE: Instructors expect students to hand in their work on time; however, illness and personal emergencies are acceptable excuses.

7. To set off transitional expressions such as *for example, as a result, in conclusion, in other words, on the contrary,* and *on the other hand*

EXAMPLE: Chuck participates in many extracurricular activities in college. As a result, he rarely gets enough sleep.

8. To set off parenthetical expressions. Parenthetical expressions are statements that provide additional information; however, they interrupt the flow of the sentence.

EXAMPLE: Fluency in a foreign language, as we all know, requires years of instruction and practice.

A comma *should not* be used in the following situations.

1. After *that,* when *that* is used in an introductory clause

EXAMPLE: The student could not believe that he lost points on his laboratory report because of a few spelling mistakes.

2. Between cumulative adjectives, which are adjectives that would not make sense if separated by the word *and* (see Item 4 in preceding list)

The semicolon. The semicolon inserts a stop between two independent clauses not joined by a coordinating conjunction (*and, but, or, nor, for, so,* or *yet*). Each independent clause (one that contains a subject and a verb) could stand alone as a sentence, but the semicolon indicates a closer relationship between the clauses than if they were written as separate sentences.

> EXAMPLE: Outstanding student-athletes use their time wisely; this trait makes them highly sought after by employers.

A semicolon is also used to separate items in a series in which the items are already separated by commas.

> EXAMPLE: Participating in sports has many advantages. First, you are doing something good for your health; second, you enjoy the camaraderie of people with a common interest; third, you learn discipline, which helps you make effective use of your time.

The colon. The colon is used to call attention to the words that follow it. Some conventional uses of a colon are shown in the following examples.

Dear Sir or Madam:
5:30 P.M.
2:1 (ratio)

In references, to separate the place of publication and the publisher, as in

Sunderland, MA: Sinauer Associates, Inc.

A colon is often used to set off a list, as in the following example.

> EXAMPLE: Catalase activity has been found in the following vegetables: potatoes, leeks, parsnips, onions, zucchini, carrots, and broccoli.

A colon *should not* be used when the list follows the words *are, consist of, such as, including,* or *for example.*

> EXAMPLE: Catalase activity has been found in vegetables such as potatoes, leeks, parsnips, onions, zucchini, carrots, and broccoli.

The period. The period is used to end all sentences except questions and exclamations. It is also used in some abbreviations, for example, *Mr., Ms., Dr., Ph.D., i.e.,* and *e.g.*

Parentheses. Parentheses are used mainly in two situations in scientific writing: to enclose supplemental material and to enclose references to visuals or sources. Use parentheses sparingly because they interrupt the flow of the sentence.

> EXAMPLE: Human error (failure to make the solutions correctly, arithmetic errors, and failure to zero the spectrophotometer) was the main reason for the unexpected results.

> REFERENCE
> TO VISUAL: There was no catalase activity above 70°C (Figure 1).

> CITATION-
> SEQUENCE
> SYSTEM: C-fern spores do not germinate in the dark (1).

> NAME-YEAR
> SYSTEM: C-fern spores do not germinate in the dark (Cooke and others, 1987).

The dash. The dash is used to set off material that requires special emphasis. To make a dash on the computer, type two hyphens without a space before, after or in between. In some word processing programs, the two hyphens are automatically converted to a solid dash.

Similar to commas and parentheses, a pair of dashes may be used to set off supplemental material.

> EXAMPLE: Human error--failure to make the solutions correctly, arithmetic errors, and failure to zero the spectrophotometer--was the main reason for the unexpected results. (If the word processing program has been set up to convert the two hyphens to a solid dash, the sentence looks like this: Human error—failure to make…spectrophotometer—was the main reason…)

Similar to a colon, a single dash calls attention to the information that follows it.

> EXAMPLE: Catalase activity has been found in many vegetables—potatoes, leeks, parsnips, onions, zucchini, carrots, and broccoli.

If an abrupt or dramatic interruption is desired, use a dash. If the writing is more formal or the interruption should be less conspicuous, use one of the other three punctuation marks. However, do not replace a pair of dashes with commas when the material to be set off already contains commas, as in the following example.

> EXAMPLE: The instruments that she plays‸oboe, guitar, and piano‸are not traditionally used in the marching band.

Clarity

The main reason for writing a scientific paper or laboratory report is to communicate information to your peers. If you do not present this information clearly, your readers will not understand what you mean. For student writers whose papers are evaluated by instructor readers, lack of clarity translates into a low grade. Researchers and faculty members, whose reputation depends on the number and quality of publications, simply cannot afford *not* to write clearly, because poorly written papers may be equated with shoddy scientific methods.

There are a number of ways to improve the clarity of your writing. These include eliminating wordiness, redundancy, empty phrases, and ambiguity; reducing complexity; and ensuring good flow between sentences and paragraphs.

Eliminate wordiness. "Wordiness" means using too many words to convey an idea. One form of wordiness is **redundancy**, using two or more words that mean the same thing (Table 5.2).

Redundancy is easily corrected by eliminating one of the redundant words.

Empty phrases are another source of wordiness. Consider the following two sentences (inspired by Van Alstyne, 1986). Which one would you rather read?

TABLE 5.2 Examples of redundancy

REDUNDANT	REVISED
It is absolutely essential…	It is essential…
mutual cooperation	cooperation
basic fundamental concept	basic concept or fundamental concept
totally unique	unique
The solution was obtained and transferred…	The solution was transferred…

> EXAMPLE 1: It is absolutely essential that you use a mini-
> mum number of words in view of the fact that
> your reader has numerous other tasks to com-
> plete at the present time.

> EXAMPLE 2: Write concisely, because your reader is busy.

Replace empty phrases with a concise alternative (Table 5.3).
   A third source of wordiness is **needlessly complex sentences**. Watch out for the sentence constructions illustrated by the following examples.

> EXAMPLE 1: *There are* two protein assays *that* are often used
> in research laboratories.

TABLE 5.3 Examples of empty phrases

EMPTY	CONCISE
a downward trend	a decrease
a great deal of	much higher
a majority of	most
accounted for the fact that	because
as a result	so, therefore
as a result of	because
as soon as	when
at which time	when
at all times	always
at a much greater rate than	faster
at the present time, at this time	now, currently
based on the fact that	because
brief in duration	short, quick
by means of	by
came to the conclusion	concluded
despite the fact that, in spite of the fact that	although, though
due to the fact that, in view of the fact that	because
for this reason	so

REVISION: Two protein assays are often used in research laboratories. [Avoid expletives.]

EXAMPLE 2: *It is interesting to note that* some enzymes are stable at temperatures above 60°C.

REVISION: Some enzymes are stable at temperatures above 60°C. [Avoid unnecessary introductions.]

EXAMPLE 3: *The analyses were done on* the recombinant DNA to determine which piece of foreign DNA was inserted into the vector.

TABLE 5.3 *Continued*

EMPTY	CONCISE
in fact	*omit*
functions to, serves to	*omit*
degree of	higher, more
in a manner similar to	like
in the amount of	of
in the vicinity of	near, around
is dependent upon	depends on
is situated in	is in
it is interesting to note that, it is worth pointing out that	*omit these kinds of unnecessary introductions*
it is recommended	I (we) recommend
on account of	because, due to
prior to	before
provided that	if
referred to as	called
so as to	to
through the use of	by, with
with regard to	on, about
with the exception of	except
with the result that	so that

REVISION: The recombinant DNA was analyzed to determine which piece of foreign DNA was inserted into the vector. [Make *DNA*, not the *analyses*, the subject of the sentence.]

EXAMPLE 4: *We make the recommendation* that micropipettors be used to measure volumes less than 1 mL.

REVISION: We recommend that micropipettors be used to measure volumes less than 1 mL. [Replace sluggish noun phrases with verb phrases.]

EXAMPLE 5: These assays alone cannot tell what the protein concentration of a substance is.

REVISION: These assays alone cannot determine the protein concentration of a substance. [Replace colloquial expressions with precise alternatives.]

Eliminate ambiguity. Avoid vague use of *this, that,* and *which* when referring to topics mentioned previously. What does the writer mean by *this* and *which* in the following example?

EXAMPLE: The data show that the longer the enzyme was exposed to the salt solution, the lower the enzyme activity in the assay. *This* means that the salt changes the conformation of the enzyme, *which* makes it less reactive with the substrate.

The subsequent revision eliminates the ambiguity.

REVISION: The data show that the longer the enzyme was exposed to the salt solution, the lower the enzyme activity in the assay. Exposure to the salt solution may change the conformation of the enzyme, resulting in lower enzyme-substrate activity.

Another source of ambiguity is when a pronoun (him, her, it, he, she, its) could refer to two possible antecedents. In the following example, can you determine what the writer meant by *it*?

EXAMPLE: With time, salt changes the conformation of the enzyme, which makes *it* less reactive with the substrate.

If there is any doubt about what *it* refers to, replace *it* with the appropriate noun phrase.

REVISION: With time, salt changes the conformation of the enzyme, so that the enzyme can no longer react with its substrate.

Use connecting words and repetition to improve flow. A smooth transition from one sentence to the next is essential for reader comprehension. Consider the following example:

EXAMPLE: Catalase is an enzyme that breaks down hydrogen peroxide in both plant and animal cells. Low or high temperature can lower the rate at which the catalase can react with the hydrogen peroxide. In optimal conditions, the enzyme functions at a rate that will prevent any substantial buildup of the toxin. If the temperature is too low, the rate will be too slow, but high temperatures lead to the denaturation of the enzyme.

Where is the writer going with this paragraph? The sentences do not seem to flow, because there is no guidance from the writer on how one sentence is related to the next. To improve flow, use connecting words such as *however, thus, although, in contrast, similarly, on the other hand, in addition to,* and *furthermore.* Repetition of a key word from the previous sentence also helps the reader make connections between sentences.

REVISION: Catalase is an enzyme that breaks down hydrogen peroxide in both plant and animal cells. One of the factors that affects the rate *of this reaction* is temperature. At optimal *temperatures,* the rate is sufficient to prevent substantial buildup of the toxic hydrogen peroxide. If the temperature is too low, *however,* the rate will be too slow, and hydrogen peroxide *accumulates* in the cell. *On the other hand,* high temperatures may denature the enzyme.

Grammar

Grammar refers to the rules that deal with the form and structure of words and their arrangement in sentences. This section describes three of the most common offenses. See Hacker (1997), Lunsford (2001), or Lunsford and Connors (1995, 1999) for a more comprehensive treatment of the subject.

Make subjects and verbs agree. We learn early on in our formal education to make the verb agree with the subject. Most of us know that *the sample was…*, but that *the samples were…* Most errors with subject-verb agreement occur when there are words *between* the subject and the verb, as in the following example.

> EXAMPLE: *The samples* in the assay *were* [not was] incubated
>   at 37°C for 10 min.

When you write complex sentences, ask yourself what is the subject of the sentence. Look for the verb that goes with that subject, and then mentally remove the words in between. Make the subject and its verb agree.

A second situation in which subject–verb agreement becomes confusing is when there are two subjects joined by *and,* as in the following example.

> EXAMPLE: An enzyme's amino acid *sequence and* its three-
>   dimensional *structure make* [not makes] the
>   enzyme-substrate relationship unique.

Compound subjects joined by *and* are almost always plural.

A third situation involving subject–verb agreement is that when numbers are used in conjunction with units, the quantity is considered to be singular, not plural.

> EXAMPLE: To extract the enzyme, 50 g of potatoes *was* [not were]
>   homogenized with 50 mL of cold, distilled water.

Write in complete sentences. A complete sentence consists of a subject and a verb. If the sentence starts with a subordinate word or words such as *after, although, because, before, but, if, so that, that, though, unless, until, when, where, who,* or *which,* however, another clause must complete the sentence.

> EXAMPLE 1: High temperatures destroy the three-dimen-
>   sional structure of enzymes. Thus changing
>   the effectiveness of the enzymes. [The second
>   "sentence" is a fragment.]

> REVISION: High temperatures destroy the three-dimen-
>   sional structure of enzymes, thus changing
>   their effectiveness. [Combine the fragment
>   with the previous sentence, changing punctua-
>   tion as needed.]

> EXAMPLE 2: The standard curve for the Biuret assay was
>   used to determine the protein concentration of
>   the serial dilutions of the egg white. Although
>   only those dilutions whose protein concentra-

tions fell within the sensitivity range of the assay were multiplied by the dilution factor to give the original concentration of the egg white. [The second "sentence" is a fragment.]

REVISION: The standard curve for the Biuret assay was used to determine the protein concentration of the serial dilutions of the egg white. Only those dilutions whose protein concentrations fell within the sensitivity range of the assay were multiplied by the dilution factor to give the original concentration of the egg white. [Delete the subordinate word(s) to make a complete sentence.]

Revise run-on sentences. Run-on sentences consist of two or more independent clauses joined without proper punctuation. Each independent clause could stand alone as a complete sentence. Run-on sentences are common in first drafts, where your main objective is to get your ideas down on paper (or electronic media, if you use a computer). When you revise your first draft, however, use one of the following strategies to revise run-on sentences:

- Insert a comma and a coordinating conjunction (*and, but, or, nor, for, so,* or *yet*).
- Use a semicolon or possibly a colon.
- Make two separate sentences.
- Rewrite the sentence.

EXAMPLE 1: The class data for the Bradford method were scattered, those for the Biuret method were closer.

REVISION A: The class data for the Bradford method were scattered, but those for the Biuret method were closer. [Use a coordinating conjunction.]

REVISION B: The class data for the Bradford method were scattered; those for the Biuret method were closer. [Use a semicolon.]

EXAMPLE 2:   The readings from the spectrophotometer should show a correlation between protein concentration and absorbance, this is Beer's law, which relates absorbance to the path length of light along with molar concentration of a solute and the molar coefficient. [Fused sentence]

REVISION A: The readings from the spectrophotometer should show a correlation between protein concentration and absorbance; this is Beer's law, which relates absorbance to the path length of light along with molar concentration of a solute and the molar coefficient. [Use a semicolon to separate the two clauses.]

REVISION B: The readings from the spectrophotometer should show a correlation between protein concentration and absorbance. This relationship is described by Beer's law, which relates absorbance to the path length of light along with molar concentration of a solute and the molar coefficient. [Make two separate sentences.]

EXAMPLE 3:   An increase in enzyme concentration increased the reaction rate as did an increase in substrate concentration, so the concentrations of the molecules have an influence on how the enzyme reacts.

REVISION A: As enzyme concentration and substrate concentration increased, so did the reaction rate. [Rewrite the sentence. The second half of the original sentence was deleted because it is deadwood.]

REVISION B: Enzyme and substrate concentration influence enzyme reaction rate: an increase in enzyme or substrate concentration increased reaction rate. [Use a colon.]

Word usage

When you write the right words in the right situations, readers have confidence in your work. Use a standard dictionary whenever you are not sure about word usage. Consult your textbook and laboratory exercise for proper spelling and usage of technical terms. The following word pairs are frequently confused in students' biology papers.

absorbance, absorbency, observance   *Absorbance* is how much light a solution absorbs; absorbance is measured with a spectrophotometer. *According to Beer's law, absorbance is proportional to concentration. Absorbency* is how much moisture a diaper or paper towel can hold. *Brand A paper towels show greater absorbency than Brand B paper towels. Observance* is the act of observing. *Government offices are closed today in observance of Independence Day.*

affect, effect   *Affect* is a verb that means "to influence." Affect is never used as a noun. *Temperature affects enzyme activity. Effect* can be used either as a noun or a verb. When used as a noun, *effect* means "result." *We studied the effect of temperature on enzyme activity.* When used as a verb, effect means "to cause." *High temperature effected a change in enzyme conformation, which destroyed enzyme activity.*

alga, algae   See Plurals.

amount, number   Use *amount* when the quantity cannot be counted. *The reaction rate depends on the amount of enzyme in the solution.* Use *number* if you can count individual pieces. *The reaction rate depends on the number of enzyme molecules in the solution.*

analysis, analyses   See Plurals.

bacterium, bacteria   See Plurals.

bind, bond   *Bind* is a verb meaning "to link." *The active site is the region of an enzyme where a substrate binds. Bond* is a noun that refers to the chemical linkage between atoms. *Proteins consist of amino acids joined by peptide bonds. Bond* used as a verb means "to stick together." *This 5-minute epoxy glue can be used to bond hard plastic.*

complementary, complimentary   *Complementary* means "something needed to complete;" matching. *The DNA double helix consists of complementary base pairs: A always pairs with T, and G with C. Complimentary* means "given free as a courtesy." *The brochures at the Visitor's Center are complimentary.*

confirmation, conformation   *Confirmation* means "verification." *I received confirmation from the postal service that my package had arrived. Conformation* is the three-dimensional structure of a macromolecule. *Noncovalent bonds help maintain a protein's stable conformation.*

continual, continuous   *Continual* means "going on repeatedly and frequently over a period of time." *The continual chatter of a group of inconsiderate students during the lecture annoyed me. Continuous* means "going on without interruption over a period of time." *The bacteria were grown in L-broth continuously for 48 hr.*

create, prepare, produce   *Create* is to cause to come into existence. *The artist used wood and plastic to create this sculpture. Prepare* means "to make ready." *The protein standards were prepared from a 50 mg/mL stock solution. Produce* means to make or manufacture. *The reaction between hydrogen peroxide and catalase produces water and oxygen.*

datum, data   See Plurals.

different, differing   *Different* is an adjective that means "not alike." An adjective modifies a noun. *Different concentrations of bovine serum albumin were prepared. Differing* is the intransitive tense of "to differ," a verb that means "to vary." It is incorrect to replace the word *different* with *differing* in the preceding example, because *differing* implies that a single concentration changes depending on time or circumstance. This situation is highly unlikely with bovine serum albumin, which is quite stable under laboratory conditions! An acceptable use of *differing* is shown in the following example. *Bovine serum albumin solutions, differing in their protein content, were prepared.*

effect, affect   See affect, effect.

fewer, less   Use *fewer* when the quantity can be counted. *The reaction rate was lower, because there were fewer collisions between enzyme and substrate molecules.* Use *less* when the quantity cannot be counted. *The weight of this sample was less than I expected.*

formula, formulas, formulae   See Plurals.

hypothesis, hypotheses   See Plurals.

its, it's   *Its* is a possessive pronoun meaning "belonging to it." *The Bradford assay is preferred because of its greater sensitivity. It's* is a contraction of "it is." *The Bradford assay is preferred because it's more sensitive.* (*Note:* Contractions should not be used in formal writing.)

less, fewer   See fewer, less

lose, loose   *Lose* means to misplace or fail to maintain something. *An enzyme may lose its effectiveness at high temperatures. Loose* means "not tight." *When you autoclave solutions, make sure the lid on the bottle is loose.*

lowered, raised   Both of these are transitive verbs, which means that they require a direct object (a noun to act on). **Wrong**: *The fish's body temperature lowered in response to the cold water.* **Right**: *The cold water lowered the fish's body temperature.*

media, medium   See Plurals.

observance   See absorbance, absorbency, observance.

phenomenon, phenomena   See Plurals.

Plurals   The plural and singular forms of some words used in biology are given in Table 5.4. A common mistake with these words is failure to make the subject and verb agree. Some disciplines treat *data* as singular, but scientists and engineers subscribe to the strict interpretation that *data* is plural. The data *show*... [not *shows*] is correct.

TABLE 5.4 Singular and plural of some common biological words

SINGULAR	PLURAL
alga	algae
analysis	analyses
bacterium •	bacteria
criterion	criteria
datum (rarely used)	data
formula	formulas, formulae
hypothesis	hypotheses
index	indexes, indices
medium	media
phenomenon	phenomena
ratio	ratios

prepare   See create, prepare, produce.

produce   See create, prepare, produce.

raised, lowered   See lowered, raised.

ratio, ration   *Ratio* is a proportion or quotient. *The ratio of protein in the final dilution was 1:5. Ration* is a fixed portion, often referring to food. *The Red Cross distributed rations to the refugees.*

strain, strand   A *strain* is a line of individuals of a certain species, usually distinguished by some special characteristic. *The lacI⁻ strain of E. coli produces a nonfunctional repressor protein.* A *strand* is a ropelike length of something. *The strands of DNA are held together with hydrogen bonds.*

that, which   Use *that* with restrictive clauses. A restrictive clause limits the reference to a certain group. Use *which* with nonrestrictive clauses. A nonrestrictive clause does not limit the reference, but rather provides additional information. Commas are used to set off nonrestrictive clauses but not restrictive clauses. Consider the following examples:

EXAMPLE 1: The Bradford assay, which is one method for measuring protein concentration, requires only a small amount of sample. [*Which* begins a phrase that provides additional information, but is not essential to make a complete sentence.]

EXAMPLE 2: Enzyme activity decreased significantly, which suggests that the enzyme was denatured at 50°C. [*Which* refers to the entire phrase *Enzyme activity decreased significantly,* not to any specific element.]

EXAMPLE 3: The samples that had high absorbance readings were diluted. [*That* refers specifically to *The samples.*]

than, then   *Than* is an expression used to compare two things. *Collisions between molecules occur more frequently at high tempera-tures than at low temperatures. Then* means "next in time." *First 1 mL of protein sample was added to the text tube. Then 4 mL of Biuret reagent was added.*

various, varying   *Various* is an adjective that means "differ-ent." *Various hypotheses were proposed to explain the observations. Varying* is a verb that means "changing." *Varying the substrate concentration while keeping the enzyme concentration constant allows you to determine the effect of substrate concentration on enzyme activity.* Analogous to *different, differing,* replacing the word *various* with *varying* in the preceding example changes the meaning of the sentence. *Varying* implies that a single hypothe-sis changes depending on time or circumstance. *Various* implies that different hypotheses were proposed.

Spelling

Spell checkers in word processing programs are so easy to use that there is really no excuse for *not* using them. Just remember that spell checkers may not know scientific terminology, so consult your textbook or laboratory manual for correct spelling. In some cases, the spell checker may even try to get you to change a properly used scientific word to an inappropriate word that happens to be in its database (for example, *absorbance* to *absorbency*).

The following poem is an example of how indiscriminate use of the spell checker can produce garbage:

Wrest a Spell

Eye halve a spelling chequer
It came with my pea sea
It plainly marques four my revue
Miss steaks eye kin knot sea.

Eye strike a key and type a word
And weight four it two say
Weather eye am wrong oar write
It shows me strait a weigh.

As soon as a mist ache is maid
It nose bee fore two long
And eye can put the error rite
Its rare lea ever wrong.

Eye have run this poem threw it
I am shore your pleased two no
Its letter perfect awl the weigh
My chequer tolled me sew.

— Sauce unknown

Spell checkers will also not catch mistakes of usage, for example *form* if you really meant *from*. Print out your document and proofread the hard copy carefully.

Global Revision

Organization. Is your document formatted according to "Instructions to Authors"? Is your scientific paper divided into sections? Does each section have the right content? The "Biology Lab Report Checklist" (pp. 96–97) and "Laboratory Report Mistakes" (pp. 108–111) tables can alert you to potential problem areas.

Content. Is your paper focussed on the subject? Did you provide all of the necessary supporting materials (tables and figures) to support your conclusions? Did you answer all of the questions in the laboratory exercise? Use the "Biology Lab Report Checklist" and "Laboratory Report Mistakes" key.

Style. Read your paper aloud. Listen for awkward sentence structure and illogical reasoning. Avoid wordiness and flowery language that might make your meaning ambiguous. Use definite and specific sentences. Keep related words together. Consider the following sentence taken from an English-language newspaper in Japan: "A committee was formed to examine brain death in the Prime Minister's office." Although brain death in the Prime Minister's office may be a political reality, what was really intended was "A committee was formed in the Prime Minister's office to examine brain death."

Focus on the following to achieve a clear and readable style.

1. Paragraphs
   - Each paragraph focuses on one topic.
   - First sentence introduces the topic.
   - Subsequent sentences support the topic sentence.
   - Connecting phrases are used to achieve good flow between sentences.

2. Sentences
   - Eliminate wordiness.
   - Vary sentence structure and length to avoid monotony.
   - Use punctuation correctly.
   - Use passive and active voice appropriately.
   - Use past and present tense appropriately.

3. Words and phrases
   - Make them concise and descriptive.
   - Use scientific words when appropriate. Define terms that might be unfamiliar to your audience. Avoid jargon and anthropomorphism.
   - Watch out for commonly confused word pairs such as *effect* and *affect*, *it's* and *its*, *then* and *than*, and others.
   - Avoid clichés, slang, and sexist language.
   - Do not use contractions in formal writing.

This **Biology Lab Report Checklist** can be printed from <http://www. sinauer. com/Knisely>.

TITLE (p. 57)

☐ Descriptive and concise

AUTHORS

☐ Writer is first author, lab partner(s) second (third, etc.) author

ABSTRACT (pp. 56–57)

☐ Contains purpose of experiment
☐ Contains brief description of methods
☐ Contains results
☐ Contains conclusions

INTRODUCTION (pp. 55–56)

☐ Contains background information from the literature (primary references)
☐ Selected references are directly relevant to your experiment.
☐ Citation format is correct.
☐ Citations are paraphrased. Direct quotations are not used.
☐ Objectives of experiment are clearly stated.

MATERIALS AND METHODS

☐ Materials are not listed separately (p. 35).
☐ No references to containers (pp. 35–36)
☐ No description of routine procedures (p. 36)
☐ No preview of how the data will be organized or interpreted (p. 37)
☐ Written in paragraph form (not a numbered list) (p. 35)
☐ Written in past tense (p. 35)
☐ Contains all relevant information to enable the reader to repeat the experiment. This includes volumes, temperatures, wavelengths, and so on (p. 37).

Biology Lab Report Checklist

RESULTS

- ☐ Figures and tables are present. Figure captions go below figure; table captions go above table (pp. 42–48)
- ☐ Text is present. Reference is made to each table and figure, and the results are described in words (pp. 39–41).
- ☐ Figure and table titles are informative and can be understood apart from the text (pp. 42–44).
- ☐ No explanation is given for the results (p. 38).

DISCUSSION (p. 55)

- ☐ Results are briefly restated.
- ☐ Explanations for results are given; implications are considered.
- ☐ Errors and inconsistencies are pointed out.

REFERENCES (pp. 57–65)

- ☐ References consist mostly of primary journal articles, not textbooks or Internet sources.
- ☐ Reference format is correct and complete.
- ☐ All references have been cited in the text. All citations in the text have been included in the References section.

REVISION

- ☐ Conventions in biology (pp. 71–73)
- ☐ Numbers (p. 74)
- ☐ Standard abbreviations (pp. 75, 76–77)
- ☐ Punctuation (pp. 75, 78–81)
- ☐ Clarity (pp. 81–85)
- ☐ Grammar (pp. 85–88)
- ☐ Word usage (pp. 89–93)
- ☐ Spelling (pp. 93–94)
- ☐ Global revision (pp. 94–95)
- ☐ All questions from the laboratory exercise have been answered

Biology Lab Report Checklist

Chapter 6

A "Good" Sample Student Laboratory Report

The laboratory report in this chapter was written by Lynne Waldman during her first year at Bucknell University, in an introductory course for biology majors. Lynne and her lab partners designed and carried out an original project in which they investigated the effect of a fungus on the growth of bean, pea, and corn plants.

Lynne's report has many of the characteristics of a well-written scientific paper. When you look over her presentation, notice the style and tone of her writing, as well as the format of the paper. The comments and annotations alert you to important points to keep in mind when you write your laboratory report.

The presentation here has been typeset to fit this book and to accommodate the annotations. Your report should be formatted to fit standard $8\frac{1}{2} \times 11$ inch paper. Unless you are instructed otherwise, use a serif type (Times Roman is standard), double space, and leave *at least* 1 inch of margin all around.

For details on how to format documents in Microsoft Word, see "Formatting a Document" in Appendix 1.

The Effects of the Fungus *Phytophthora infestans* on Bean, Pea, and Corn Plants

Lynne Waldman, Partner One, Partner Two

Informative title

Author's name first, followed by lab partners' names in alphabetical order

Sections of report are clearly labeled.

Background information is present.

No references are used in Abstract.

Purpose of current experiment is given.

Short description of methods

Results are given.

No references are made to figures.

Short explanation of results is given.

Abstract is a maximum of 250 words long.

Background information

Latin names are italicized.

Abstract

Phytophthora infestans is a fast-spreading, parasitic fungus that caused the infamous potato blight by devastating Ireland's crops in the 1840s. *P. infestans* also causes late blight in tomato plants, a relative of the potato. In this experiment, the effects of *P. infestans* on *Phaseolus* variety long bush bean, *Zea mays* (corn), and *Pisum sativum* (pea) were studied. The soil surrounding the roots of 18-day old plants was injected with *P. infestans* cultured in an L-broth medium. Plant height, number of leaves, and leaf angle were measured for each plant during the next 8 days. Chlorophyll assays were performed prior to exposure, and on the eighth day after exposure to the fungus. The plants were also examined for black or brown leaf spots characteristic of late blight infections. The results showed that *P. infestans* had no apparent effect on the bean, corn, and pea plants. One reason for this may be that there were no fungus zoospores in the L-broth medium. More probably, however, *P. infestans* may be a species-specific pathogen that cannot infect bean, corn, or pea plants.

Introduction

Originating in Peruvian-Bolivian Andes, the potato *(Solanum tuberosum)* is one of the world's four most important food crops (along with wheat, rice, and corn). Cultivation of potatoes began in South America over 1,800 years ago, and through the Spanish conquistadors, the tuber was introduced into Europe in the second half of the 1600s. By the beginning of the 18th century, the potato was widely grown in Ireland, and the country's economy heavily relied on the potato crop. In the middle of the 19th century, Ireland's potato crop suffered widespread late blight disease caused by *Phytophthora infestans,* a species of pathogenic plant fungus. Failure of the potato crop because of late blight resulted in the Irish potato famine. The famine led to widespread starvation and the death of about a million Irish.

The potato continues to be one of the world's main food crops. However, *P. infestans* has reemerged in a chemical-resistant form in the United States, Canada, Mexico, and Europe (McElreath, 1994). Late blight caused by the new strains is costing growers worldwide about $3 billion annually. The need to apply chemical fungicides eight to ten times a season further increases the cost to the grower (Stanley, 1994 and Stanley, 1997). *P. infestans* is thus an economically important pathogen.

P. infestans, which can destroy a potato crop in the field or in storage, thrives in warm, damp weather. The parasitic fungus causes black or purple lesions on a potato plant's stem and leaves. As a result of infection by this fungus, the plant is unable to photosynthesize, develops a slimy rot, and dies. *P. infestans* similarly infects the tomato plant *(Lycopersicon esculentum)* (Anonymous, 1994).

The purpose of the present experiment was to determine the effects of *P. infestans* on plant height, number of leaves, leaf angle, and chlorophyll content of three agriculturally important plants: *Phaseolus* variety long bush beans, *Zea mays* (corn), and *Pisum sativum* (peas). Symptoms of fungal infection were assumed to be similar to that in potatoes.

Materials and Methods

Phaseolus variety long bush bean, *Zea mays* (corn) and *Pisum sativum* (pea) seeds were soaked overnight in tap water. Fifteen randomly chosen seeds of each species were planted 1 cm beneath the surface in three separate trays containing 10 cm of potting soil. Another set of trays, which was to be the control group, was prepared in the same fashion.

Margin notes:

Proper citation format is used (Name-Year system).

No direct quotations are used. Citations are paraphrased and the source is given in parentheses.

"Anonymous" is used when no author is given in newspaper or magazine articles.

Purpose of experiment is clearly stated.

M&M are always written in past tense.

Sufficient detail is given to allow the reader to repeat the experiment.

All the experimental plants were placed in one fume hood, and all the control plants were placed in relative positions in another fume hood in the same room. The plants were exposed to the ambient light intensity in the hood (153 fc) and air current 24 hrs a day, and were watered lightly daily. The plants were allowed to germinate and grow for 18 days.

Phytophthora infestans on potato dextrose agar was obtained from Carolina Biological Supply House. At day 10 of the plant growth regime, pieces of agar on which the fungus was growing were transferred to L-broth. L-broth consisted of 5 g yeast extract, 10 g tryptone, 1 g dextrose, and 10 g NaCl dissolved in distilled water, and adjusted to pH 7.1, to make 1 L of medium. The medium was sterilized before adding the fungal culture. After 4 days in L-broth, 6 mL of the fungal culture was injected into the soil around the roots of each 18-day old plant. 6 mL of L-broth without *P. infestans* was injected into the soil of the control plants. All plants were then allowed to grow for another 8 days.

Every other day after treatment with *P. infestans*, plant height and number of leaves were measured for both the control and the experimental plants. Plant height was measured from the soil to the apical meristem of the plant. Leaf angle (as shown in Figure 1) of the largest, lowest leaf on each plant was measured three times, once prior to injection, once 4 days after injection, and once 8 days after injection. Leaf angle was measured in order to

Figures that explain the methodology may be included in M&M section.

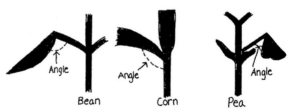

Figure 1 Leaf angle as measured in bean, corn, and pea plants

determine if *P. infestans* causes wilting in the three plant species. In addition, the plant was examined visually for the presence of any leaf spots.

Chlorophyll assays were performed on one plant from each tray prior to injection and on the eighth day after injection. For each chlorophyll assay, the leaves of the plant were removed from the stem. For each 0.1 g of leaves, 6.0 mL of 100% methanol were used. The leaves were thoroughly ground in half of the methanol with a pestle in a mortar. The leaves were ground again after the rest of the methanol was added. Extraction of the chlorophyll was allowed to proceed for 45 min at room temperature. Then the suspension was gravity filtered through filter paper to remove the leaf parts. The absorbance of the filtrate was measured with a Spectronic 20 spectrophotometer at 652 nm and 665.2 nm. The absorbance values were converted to relative chlorophyll units using the following equation derived by Porra and colleagues (1989):

Total chlorophyll (a and b) = Dilution factor × [22.12 $A_{652\,nm}$ + 2.71$A_{665.2\,nm}$ (mg/L)] × Volume of solvent (L) / Weight of leaves (mg)

Subscripts are made properly.

Results

P. infestans-treated plants and the control plants had similar growth patterns (Figure 2). Both the experimental and control pea and corn plants grew at a

Results section must contain a text in which the author presents each figure and table to the reader.

Reference is made to the next figure in the sequence, and the important results are described.

Figure is large enough to read key and axes easily.

Axes have proper spacing.

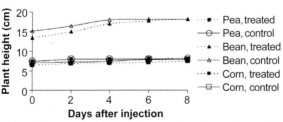

Figure 2 Average height of control and experimental plants in the period after injection with *P. infestans*

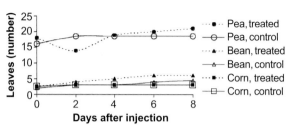

Figure 3 Average number of leaves of control and experimental plants in the period after injection with *P. infestans*

Points and lines on curve are dark and can be easily distinguished from each other.

Units are given in parentheses after the axis label (where applicable).

Figure caption is located below figure.

Figure title is not simply a repeat of "*y*-axis label vs. *x*-axis label ."

Figure number sequence is correct.

Reference is made to the next figure and the important information is described.

Symbols (such as °) are typed using word processing software.

constant, but very slow rate over the eight day test period. The control bean plants were taller on average than the experimental bean plants throughout most of the experiment. Both groups showed the same growth pattern, however, with rapid growth occurring from day 18 to 24 (0 to 4 days after injection), followed by slower growth to the end of the experiment.

As plant height increased, the average number of leaves on all of the plants also increased over the measurement period (Figure 3). There is an uncharacteristic decrease in the number of leaves of pea plants treated with *P. infestans* from day 18 to 20 (0 to 2 days after injection), but this is probably due to counting error.

There was a general decline in average leaf angle of all the plants over the first four days after injection with *P. infestans* (Figure 4). The plants did not follow this pattern over the second half of the experiment, however. The leaf angle of the experimental bean group increased by 28°, while that of the control bean group only increased by about 3°. The leaf angle of the control pea plants increased significantly (33°), while that of the experimental pea plants decreased 4°. The leaf angle of the corn control group decreased 0.5°, while that of the corn experimental group showed a much sharper decline of 24°.

There was also no difference between the experimental and control groups with regard to chloro-

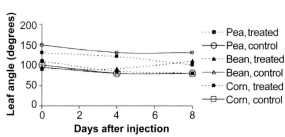

Figure 4 Average leaf angle of control and experimental plants in the period after injection with *P. infestans*

phyll content. There was a slight increase in chlorophyll content from day 18 to 26 (0 to 8 days after injection) in the corn plants (Table 1). For the bean group, there was a large decrease in chlorophyll content, 0.1 relative chlorophyll units, which did not seem to agree with the general appearance of the plants. There may have been some error when this assay was carried out. There was little change in chlorophyll content for the pea group.

Finally, there was no evidence of any brown or black leaf spots symptomatic of *P. infestans* infection.

Table is numbered properly.

Table 1 Chlorophyll content of corn, bean, and pea plants prior to infection and 8 days after infection

Table caption is placed above the table.

Plant	Relative chlorophyll units		Change in chlorophyll content (relative units)
Plant	Day 0	Day 8	
Corn, treated	9.036×10^{-4}	9.383×10^{-4}	$+3.45 \times 10^{-5}$
Corn, control	9.270×10^{-4}	8.963×10^{-4}	$+3.34 \times 10^{-5}$
Bean, treated	1.034×10^{-1}	1.2×10^{-3}	-1.022×10^{-1}
Bean, control	1.7×10^{-3}	1.6×10^{-3}	-1×10^{-4}
Pea, treated	1.3×10^{-3}	1.7×10^{-3}	$+4 \times 10^{-4}$
Pea, control	1.2×10^{-3}	1.2×10^{-3}	0.0000

Scientific notation is used correctly.

It is preferable to leave extra space at the bottom of the page, rather than to split a table across a page.

Discussion

P. infestans did not affect the plant height, leaf angle, number of leaves, and chlorophyll content of *Zea mays, Pisum sativum,* or *Phaseolus.* Symptoms of infection are the presence of brown or black spots (areas of dead tissue) on leaves and stems, and, as the infection spreads, the entire plant becomes covered with a cottony film (Stanley, 1994). None of the experimental plants exhibited these symptoms.

There may be several reasons why *P. infestans* did not affect the plants in this study. One reason is that the L-broth culture of *P. infestans* may not have contained zoospores of the fungus. Zoospores are motile spores that can penetrate the host plant through the leaves and soft shoots, or through the roots (Stanley, 1994). Zoospores are usually produced in wet, warm weather conditions (Ingold and Hudson, 1993). If the L-broth culture did not contain any zoospores, or if the soil around the plants was not sufficiently saturated to stimulate production of zoospores, then these conditions may have prevented *P. infestans* from attacking the roots and shoots of the plants.

In order to determine if the problem was lack of zoospores, first the L-broth culture could be examined microscopically for presence of zoospores. Second, the *P. infestans* plants could be watered with different quantities of water to determine if the fungus requires wetter soil for zoospore production and motility.

Another reason why *P. infestans* may not have affected the plants is that this species of fungus may be specific to potato *(Solanum tuberosum)* and tomato *(Lycopersicon esculentum)* plants (Stanley, 1994), which both belong to the nightshade family (Solanaceae). In contrast, corn belongs to the grass family (Gramineae), and peas and beans are legumes (Leguminosae). It may be that these plant families are not susceptible to *P. infestans,* which has a very limited host range (Stanley, 1994). Non-

Results are summarized or briefly restated in the Discussion section.

Explanations for results are given.

References are used to support explanations.

Ways to test explanations may be offered.

Frequent use of references is made to support explanations. Whenever possible, use primary references (journal articles, conference proceedings, collections of primary articles in a book). Avoid textbooks (secondary references) and Internet sources (may be unreliable).

susceptible plants have been shown to have defense mechanisms that prevent *P. infestans* from infecting them (Gallegly, 1995).

Further research is required to determine if *P. infestans* really cannot infect corn, pea, and bean plants. Goth and Keane (1997) developed a test to measure resistance of potato and tomato varieties to original and new strains of *P. infestans*. Their experiments involved exposing the experimental plants' leaves directly to the fungus, and this method could perhaps be tested on corn, pea, and bean leaves as well.

When Name-Year citation format is used, authors are listed alphabetically in the References section.

References

[Anonymous]. 1994. Brave New Potato. Discover 15(10): 18–20.

Gallegly ME. 1995. New criteria for classifying Phytophthora and critique of existing approaches. In: Erwin DC, Bartnicki-Garcia S, Tsao PH, editors. <u>Phytophthora: Its Biology, Taxonomy, Ecology, and Pathology</u> St. Paul: The American Phytopathological Society. pp. 167–172.

Goth RW, Keane J. 1997. A detached-leaf method to evaluate late blight resistance in potato and tomato. American Potato Journal 74(5): 347–352.

Ingold CT, Hudson HJ. 1993. <u>The Biology of Fungi</u>, 6th ed. London: Chapman and Hall.

McElreath, Linda R. 1994. One potato, two potato. Agricultural Research 42(5): 2–3.

Porra RJ., Thompson WA, Kriedemann PE. 1989. Determination of accurate extinction coefficients and simultaneous equations for assaying chlorophylls a and b extracted with four different solvents: verification of the concentration of chlorophyll standards by atomic absorption spectroscopy. Biochimica et Biophysica Acta 975: 384–394.

Stanley D. 1994. What was around comes around. Agricultural Research 42(5): 4–8.

Stanley D. 1997. Potatoes. Agricultural Research 45(5): 10–14.

Most references should be primary journal articles or articles in books. Textbooks and articles in newspapers and magazines are secondary references (less preferable).

List all authors (up to 10; then list first 10 followed by "et al." or "and others.")

See the tabbed pages in Chapter 4 for examples of how to reference journal articles, articles in books, and books.

List the starting and ending pages of the article, not just the page(s) you extracted information from.

All citations must have a corresponding reference.

All references must be cited in the text.

Laboratory Report Mistakes

The following table lists some common mistakes made by students writing biology lab reports. When you proofread the first draft of your report, look over this list and be on the lookout for these kinds of mistakes. This table can be printed from <http://www.sinauer.com/Knisely>.

Your instructor may also use this key to save time grading lab reports without sacrificing the quality of the feedback. For example, the number "8" written in the left margin means that you failed to include essential information in the Abstract. If your instructor uses this system, be sure to refer to the key for an explanation of your instructor's comments.

Although some of these mistakes may not affect the content of your paper, they do affect your credibility as a scientist; careless writing may be equated with careless science.

MISTAKE	EXPLANATION
lc	Use lowercase letter
CAP or uc	Use capital (uppercase) letter
∧	Insert text
⌡⌠	Leave space between the two characters
∩∪	Close up
¶	Start a new paragraph
Page break	End the current page; move subsequent text to next page. (In Microsoft Word and WordPerfect, press Ctrl + Enter where you want to end the page.)
agr	Subject and verb do not agree.
wc	Word choice. Word used is not appropriate for the situation.
· · · · · ·	A dotted underline means "stet," or "let original text stand." The correction was made in error.
1	Word usage incorrect. Look it up in the dictionary. Examples:
a.	absorbance (how much light is absorbed) vs. absorbency (how much moisture a diaper or paper towel can hold)
b.	to bind (what atoms and molecules do) vs. to bond (what glue does)
c.	complementary (DNA strands) vs. complimentary (given free as a courtesy)
d.	data = plural of "datum"; e.g. "These data show…" *not* "This data shows…"

e.	effect (noun) vs. effect (verb meaning "to cause") vs. affect (verb meaning "to influence")
f.	conformation (3-D shape of a protein) vs. confirmation (verification)
i.	it's (it is) vs. its (belonging to "it")
m.	amount (cannot count individual pieces) vs. number (can count individual pieces)
r.	ratio (proportion or quotient) vs. ration (a fixed portion of food)
s.	strain of bacteria vs. strand of DNA
t.	then (next in time) vs. than (an expression for comparing two things)
v.	varying (changing) vs. various (different)
2	Latin names should be *italicized*.
3	Write in passive voice. Shift the emphasis from yourself to the subject of the action.
4	Punctuation usage is incorrect.
5	Sentence structure is awkward or sentence is incomplete.
6	Symbols should not be handwritten. Use the symbol font or **Insert	Symbol** on the menu bar.
7	Superscript/Subscript. Superscript: For 10^{-5}, for example, this is done by first typing "10–5"; then highlight "–5" and either click the Superscript button on the toolbar or select **Format	Font** from the menu bar, and click **Superscript**. Subscript: For K_m, for example, this is done by first typing "Km"; then highlight "m" and either click the Subscript button on the toolbar or select **Format	Font** from the menu bar, and click **Subscript**.
8	Essential Abstract content is missing. Abstract must contain purpose, brief description of methods, results, and conclusions or possible explanation for the results.
9	Too much detail for Abstract.
10	Introduction needs references to provide background.
11	Introduction needs purpose of the current experiment.
12	Wrong citation format. Use either name-year or citation-sequence format. See Chapter 4 for specific examples.
13	Do not use direct quotations in scientific writing. Paraphrase and cite the source.

Laboratory Report Mistakes

14	Write Materials and Methods in paragraph form. Do not make a numbered list.
15	Do not list materials separately in Materials and Methods section, unless the source is noncommercial or critical for the outcome of the experiment.
16	Time frame does not affect the outcome.
17	Do not refer to the container; instead emphasize the contents and give appropriate units of concentration.
18	Do not describe routine laboratory procedures in detail.
19	In the Materials and Methods section, do not give a "preview" of how data will be organized in the Results section.
20	Essential details are missing in the Materials and Methods section. Provide enough detail to enable the reader to repeat the experiment.
21	Do not include raw data in the Results section; instead, summarize and organize the data.
22	Results section always needs a text.
23	Describe *each* table and *each* figure individually and sequentially.
24	Refer to each figure/table parenthetically by number as you describe the results.
25	A table is not needed when the figure(s) shows the same data.
26	Each figure/table must have a number and a title.
27	The figure/table title must be self-explanatory without reference to the text.
28	Uninformative figure title. Do not use "*y*-axis label vs. *x*-axis label" as a title. See Chapter 4 for examples.
29	Figure caption goes *below* figure; table caption goes *above* table.
30	No title is needed above the figure. When you make the figure in Excel, leave the "Chart title" space blank.
31	Axis does not have proper spacing. Choose "XY Scatter" plot in Excel (see Appendix 2).
32	Each data set must have its own distinguishable symbols or lines. Choose a dark color for both the line and the symbol.
33	Use scientific notation when numbers are very large or very small.

34	Units are needed.
35	Number format is incorrect. Do not start sentences with a number (write out the number). Make sure decimal numbers less than one start with zero, e.g. 0.1 mL, *not* .1 mL.
36	Discussion section requires an explanation of the results.
37	Discussion section requires a comparison of your data to what is known from the literature. Science is not done in a vacuum!
38	Wrong reference format. See Chapter 4 for specific examples.
39	All citations in the text must be listed in the References section.
40	All references must be cited in the text.

Laboratory Report Mistakes

Chapter 7

POSTER PRESENTATIONS

Posters are a means of communicating research results quickly. They provide a great opportunity to get feedback about preliminary data and ideas. **Poster sessions** are often held at large national meetings, and they allow you to meet other scientists in an informal setting.

Why Posters?

Scientists who attend poster sessions constitute a much larger audience than the one attracted to a journal article on a particular topic. Thus, your goal is to produce a poster that not only attracts experts in your subdiscipline, but also the much larger group of scientists with tangential research interests. The latter group provides a unique opportunity for you to learn about applications of your work to other research areas (and vice versa), spurs scientific creativity, and prompts you to apply an interdisciplinary approach to problem-solving.

Posters are *not* papers; they rely more on visuals than on text to present the message. It is not necessary to supply as many supporting details as you would for a paper, because you (the author) will be present to discuss details one-on-one with interested individuals. Too much material may even discourage individuals from reading your poster.

An appropriate poster presentation should fulfill two objectives. First, it must be esthetically pleasing to attract viewers in the first place. Second, it must communicate the methods, results, and conclusions clearly and concisely.

Poster Format

Size

Poster boards come in many sizes. Check with the conference organizer regarding minimum and maximum sizes. For poster sessions in your class, ask your instructor about appropriate materials and sizes.

Font (Type Size and Appearance)

Remember that most readers of your poster will be 3 to 6 feet away, so the print must be large and legible. Sans serif fonts like Arial are good for titles, but serif fonts like Times and Palatino are much easier to read in extended blocks of text. The **serifs** (small strokes that embellish the character at the top and bottom) create a strong horizontal emphasis, which helps the eye scan lines of text more easily.

Make the title Mixed Type or ALL CAPS in 72 point **bold**. Mixed type has the advantages of being easier to read and taking up less space than all caps. Do not use all caps if there are case-sensitive words in the title, such as pH, cDNA, or mRNA. Limit title lines to 65 characters or less.

Times
ARIAL

Authors' names and affiliations should be 48 or 36 point **bold**, serif font, mixed type:

Times 48 Pt

The section headings can be 28 point **bold**, serif font, mixed type:

Times 28 Pt

The text itself should be no smaller than 24 point, serif font, mixed type, and *not* in bold:

Times 24 Pt

Poster Esthetics

The success of a poster presentation depends on its ability to attract people from across the room. Interesting graphics and colored photos are good attention-getters, but avoid "cute" gimmicks. Present your poster in a serious and professional manner so people will take your conclusions seriously.

Organize the layout so that information flows from top left to bottom right. Align text on the left rather than centering it. The smooth left edge provides the reader with a strong visual guide through the material.

Avoid crowding. Large blocks of text turn off viewers; instead, use bullets to present your objectives and conclusions clearly and concisely. Use blank space to separate sections and to organize your poster for optimal flow from one section to the next.

Use appropriate graphics that communicate your data clearly. Use three-dimensional graphs *only* for three-dimensional data.

Colored borders around graphics and text enhance contrast, but keep framing to a minimum. **Framing** is the technique whereby the printed material is mounted on a piece of colored paper, which is mounted on a piece of different-colored paper to produce colorful borders. Use borders judiciously so that they do not distract from the poster content.

Nuts and Bolts

To affix text and figures to the poster board, use adhesive spray or glue stick. These tend to have fewer globs and bulges than liquid glues.

Ask the conference organizer (or your instructor) about how posters will be displayed at the poster session. Some possibilities include a pinch clamp on a pole, an easel, a table for self-standing posters, and cork bulletin boards to which posters are affixed with pushpins.

Poster Content

Posters presented at large national meetings should be organized so that readers can stand 10 feet away from the poster and get the take-home message in 30 seconds or less. Because of the large number of sessions (lectures) and an even larger number of posters, conference participants often experience "sensory overload." Thus, if you want your poster to stand out, make the section headings descriptive, the content brief and to the point, and the conclusions assertive and clear.

Posters for a student audience in the context of an in-house presentation should follow the same principle of brevity, but may retain the sections traditionally found in scientific papers. These include:

- Title banner
- Abstract (optional—if present, it is a summary of the work presented in the poster)
- Introduction
- Materials and Methods
- Results
- Discussion or Conclusions
- References (these should be brief or omitted on the poster and provided on a handout)

Title Banner

Use a short, yet descriptive, title. This is the first and most important section for attracting viewers, so try to incorporate your most important conclusion in the title. For example, **Gibberellic Acid Makes Dwarf *B. rapa* Grow Taller** is more effective than **Effect of Gibberellic Acid on Dwarf *B. rapa*.**

The title banner should be at the top of the poster and in 72 point bold font, mixed type, or all caps, 65 characters or less on a line. Underneath the title, include the authors' names, in alphabetical order, and the institutional affiliation(s). Use 48 point bold, mixed type for the authors' names.

Introduction

Instead of this conventional heading, consider using a short statement of the topic or introduce the topic as a question. Under the heading, briefly explain the existing state of knowledge of the topic, why you undertook the study, and what specifically you intended to demonstrate. A bulleted list of objectives may be a good way to present some of this information.

Materials and Methods

Present the methods you used to investigate the problem in enough detail so that someone competent in basic laboratory techniques could repeat your experiments. You might write the basic approach as a series of bulleted statements, and then provide more details in the subsequent text. Be both *brief* and *thorough*.

Results

The Results section of a poster consists mostly of visuals (pictures, tables, and graphs) and a minimum of text. Poster viewers do not have time to read the results leisurely, as they do with a journal article. An effective presentation of the results should announce each result with a heading, provide a visual that supports the result, and use text sparingly as a supplement to the visuals.

Figures are a summary of the raw data and are constructed so that viewers can appreciate both the general patterns of the data and the degree of variability that they possess. Written text should concentrate on general patterns, trends, and differences in the results, and not on the numbers themselves. For example, the reader can visualize "the concentration of chlorophyll increased initially, and then leveled off after 10 days" much more easily than "the chlorophyll concentration was 9×10^{-5} units on the fourth day, increased to 6×10^{-4} units on the tenth day, and then stayed about the same at 4×10^{-4} units on day 21."

Avoid using tables with large amounts of data; if you think the data are important, prepare the table to give as a handout. Your job is to sort through the data and come up with the take-home message. Do not use flip-out charts, in which one table or figure is displayed beneath another.

Figures may contain some statistical information including means and standard error and minimum and maximum values, where appropriate. Make the data points prominent and use a simple vertical line without crossbars for the error bars. Use the same sized font for the axis labels and the key as you use for the text (24 pt). Do not present the same data in both a table and as a figure.

Visuals in posters do not need a caption (number and title). Instead, the graphs and pictures are integrated so that they immediately follow the text in which they are first described.

Edit the text ruthlessly to remove nonessential information about the visuals. A sentence like "The effect of gibberellic acid on *B. rapa* is shown in the following figure." is nothing but deadwood. On the other hand, "*B. rapa* plants treated with gibberellic acid grew taller than those receiving only water" informs the reader of the result.

Remember to leave room for blank space on a poster. Space can be used to separate sections and gives the eyes a rest.

Discussion or Conclusions

Interpret the data in relation to the original objective or hypothesis and relate these interpretations to the present state of knowledge presented in the Introduction. Discuss any surprising results. Discuss the future needs or direction of the research. Where appropriate, identify sources of error and basic inadequacies of the technique. Do not cover up mistakes; instead, suggest ways to improve the experiment if you were to do it again. You may also speculate on the broader meaning of your conclusions in this section. Use bullets to help you present this information concisely.

Literature Citations

In scientific papers, it is common to cite the work of others, particularly in the Introduction and Discussion sections. The full references are then given in the References section at the end of the paper. Posters, on the other hand, are informal presentations that do not need to contain all the supporting details. Scientists who visit your poster are likely to work in the same field and probably are already familiar with most of the literature, so a long list of references would only waste valuable space on your poster. Even visitors who have a general knowledge of your topic but who work in a different subdiscipline are not interested in the details. They are interested mainly in how your approach or findings might help them improve their methodology or provide insight into their work. (Discussions with scientists in this second group are valuable to you because they provide a different perspective and may help you see applications to other subdisciplines.)

Student presentations in an in-house setting are different from poster sessions at large national meetings, because students usually do not have the background or the familiarity with the literature that career researchers have. Researching your topic is part of the scientific method, however, and presenting your work in the context of the published literature is part of good science. Thus, your instructor may ask you to include literature citations on your poster or to provide a list of references on a handout. See the section "Documenting Sources" in Chapter 4.

Presenting Your Poster

Authors should prepare a 5-minute talk explaining their poster; anticipate questions from students and instructors, and prepare appropriate answers. After this presentation, at least one author must be present at his or her poster at all times to answer questions from the session participants. The more you interact with your audience, the more feedback you are likely to receive on your work. (On the other hand, do not throw yourself on passersby who demonstrate little interest.)

Chapter 8

ORAL PRESENTATIONS

Scientific findings are communicated through journal articles, poster presentations at meetings, and oral presentations. Oral presentations are different from journal articles and posters because the speaker's delivery plays a critical role in the success of the communication.

The most natural form of oral presentation is extemporaneous, whereby the speaker has carefully prepared the presentation beforehand, rehearsed it, and may use an outline to stay on track during the actual talk. Most of your lectures in biology classes are likely to be extemporaneous presentations, so you may already have some experience with the characteristics that make this type of oral presentation successful:

- The speaker establishes a good rapport with the audience.

- The presentation is well organized and leaves listeners satisfied that they understand more than they did before.

- The visuals are simple, legible, and help listeners focus on the important points.

Organization

Extemporaneous presentations are not unlike scientific papers in their structure, but they are much more selective in their content. As in a scientific paper, the **introduction** captures the audience's attention, provides background information on the subject, and clarifies the objectives of the work (Table 8.1). The **body** is a condensed version of the Materials and Methods and Results sections, and contains only the details necessary to emphasize and support the speaker's conclusions. If the focus of the talk is on the results, then the speaker spends less time on the methods and more time on visuals that highlight the findings. The visuals may be simpler than those prepared for a journal article, because the

TABLE 8.1 Comparison of the structure of an oral presentation and a journal article

ORAL PRESENTATION	JOURNAL ARTICLE
	Abstract
Introduction	Introduction
Body	Materials and Methods, Results
Closing	Discussion
	References

audience may only have a minute or two to digest the material in the visual (in contrast to a journal article, where the reader can re-read the paper as often as desired). Finally, the **closing** is comparable to the Discussion section, because here the speaker summarizes the objectives and results, states conclusions, and emphasizes the take-home message for the audience.

There is no distinct Abstract or References section in an oral presentation. If the presentation is part of a meeting, an Abstract may be provided in the program. If listeners are interested in the references, they should contact the speaker after the talk.

Plan Ahead

Before you start writing the content of your presentation, make sure you know

- Your audience. What do they know? What do they want to know?
- Why you are giving the talk, not just why you did the experiment. Are you informing the audience, or are you trying to persuade them to adopt your point of view?
- Your speaking environment. How much time do you have? How large is the audience? What audio-visual equipment is available (chalkboard, overhead projector, computer with projection equipment)?

Write the Text

An oral presentation will actually take you longer to prepare than a paper. D'Arcy (1998) breaks down the steps as follows:

- Procrastination   25%
- Research   30%
- Writing and creating visuals   40%
- Rehearsal   5%

To overcome procrastination, follow the same strategy as for writing a laboratory report: write the body first (Materials and Methods and Results sections), then the introduction, and finally the closing.

- Write the main points of each section in outline form.

- Use the primary literature (journal articles) to provide background in the introduction and supporting references in the closing.

- Transfer your outline to notecards. Use only one side of the notecards, and limit yourself to one main topic per card. Note where you plan to use visual aids.

- If you plan to use a presentation graphics program such as Microsoft PowerPoint in addition to notecards, make sure each slide is just an *outline* of what you're going to say. Use the slides to *support* what you say, not *repeat* what you say.

- Keep it short and simple. Tell your audience what you're *going* to tell them, *tell* them, and then tell them what you *told* them.

Prepare the Visuals

Effective visual aids help the audience to visualize the data and to see relationships among variables. They clarify information that would otherwise involve lengthy descriptions.

Effective visuals are simple, legible, and organized logically. They capture the listener's interest, increase comprehension, and focus attention on the important findings. When deciding on which visuals to use, keep in mind that student audiences prefer figures to tables. For guidelines on what kind of figure to use, see "Visuals" and "Organizing Your Data" in Chapter 4. For instructions on how to make XY graphs in Microsoft Excel, see Appendix 2. Photos, diagrams, and drawings are effective ways to show physical rather than numerical results. Mechanisms are much easier to visualize when the verbal description is accompanied by a drawing.

Take the time to make good visuals for your oral presentation. We receive more than 83% of our information through sight, and only 11% through hearing. Without visuals, your listeners will forget most of your presentation within 8 hours. These statistics help explain why Microsoft

(A)

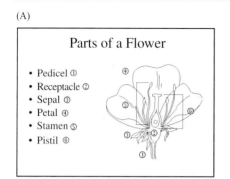

Parts of a Flower

- Pedicel ①
- Receptacle ②
- Sepal ③
- Petal ④
- Stamen ⑤
- Pistil ⑥

(B)

Parts of a Flower

- Pedicel – stalk that supports the flower
- Receptacle – enlarged tip of pedicel where floral parts attach
- Sepal – outer whorl of floral organs
- Petal – 2nd whorl of floral organs
- Stamen – 3rd whorl of floral organs; male reproductive structures
- Pistil – inner whorl of floral organs; female reproductive structures

Figure 8.1 Effective (A) and less effective (B) ways to use text on a slide. The wording in Slide A is simple and the numbered reference to the figure makes it easy for viewers to visualize the location of each floral part. The large amount of text on Slide B will have the audience reading instead of listening to the speaker, and the absence of a diagram slows comprehension.

PowerPoint has become so popular in recent years. PowerPoint allows you to create a slide show containing text, graphics, and even animations (see Appendix 3). PowerPoint has the same advantages as word processing programs in that it is easy to make revisions and even send an electronic copy of the presentation to a colleague to review.

Because PowerPoint is so easy to use, you may be tempted to write every word of your presentation on the slides. Don't do it. Follow these tips for preparing visuals:

- **Do not put too much information on any one slide or transparency.** A good rule of thumb is to cover only one concept per slide or transparency; use overlays for more complicated concepts that require more than one slide.

- **Keep the wording simple.** Instead of using full sentences, bullet keywords or key points (Figure 8.1). Limit the number of key points to 3–6.

- **Make sure the lettering can be read from anywhere in the room.** Use no smaller than 18–24 pt sans serif font and black type on a white background for best contrast. The light background makes the room much brighter, which has the dual advantage of keeping the audience awake and allowing you to see your listener's faces. Use color sparingly in text slides; a bright color or a larger font, when used sparingly, are good ways to emphasize a point.

- **Use the same font and format for each slide for a consistent look.** Avoid "cute" graphics that might detract from the scientific message.

- **Keep graphs simple.** Make the numbers and the lettering on the axis labels large and legible. Choose symbols that are easy to distinguish. Instead of following the conventions used in journal articles, modify the figure format as follows (Figure 8.2):

(A)

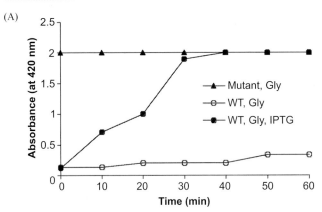

Figure 1 Beta-galactosidase production in *E. coli* grown under three different conditions. The mutant strain, which produces a nonfunctional repressor protein, was grown in glycerol. The wild-type strain was grown in glycerol, with or without the addition of IPTG (a lactose analog) at $t = 0$ min.

(B)

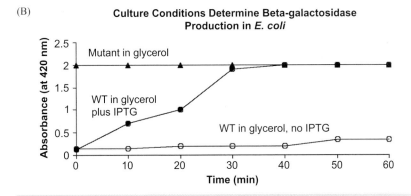

Figure 8.2   Example of figure formatting for (A) a journal article and (B) an oral presentation

- Position the title *above* the figure and *don't* number it. The title should reflect your most important finding.

- Label each line instead of identifying the data symbols with a key. The labels may be made using textboxes in Excel, Word, or PowerPoint.

Both of these modifications make it easier for the listener to digest the information within the short time the graph is displayed during the presentation.

- **Use symbols consistently in all of your graphs.** For example, if you use a triangle to represent "Mutant in glycerol" in the first graph, use the same symbol for the same condition in the rest of your graphs.

- **Appeal to your listeners' multiple senses.** Use images from the Internet, recordings of heart sounds or bird songs, and animations or movie clips if they help make your point. Make sure these "extras" do not detract from your take-home message and that they fit within the time limit of your presentation.

- **Allow at least 20 seconds for each slide, more time for complicated slides.**

Rehearsal

After you have written your presentation and prepared your visuals, you must practice your delivery. Give yourself plenty of time so that you can run through your presentation several times and, if possible, do a practice presentation in the same room where you will hold the actual presentation. Here are some other tips:

- Go to a place where you can be alone and undisturbed. Read your presentation aloud, paying attention to the organization. Does one topic flow into the next, or are their awkward transitions? Revise as needed.

- Time yourself. Make sure your presentation does not exceed the time limit.

- After you are satisfied with the organization, flow, and length, ask a friend to listen to your presentation. Ask him or her to evaluate your poise, posture, voice (clarity, volume, and rate), gestures and mannerisms, and interaction with the audience (eye contact, ability to recognize if the audience is following your talk).

It is natural to be nervous when you begin speaking to an audience, even an audience of your classmates. Adequate knowledge of the subject, good preparation, and sufficient rehearsal can all help to reduce your nervousness and enhance your self-assurance.

Delivery

The importance of the delivery cannot be overstated: listeners pay more attention to body language (50%) and voice (30%) than to the content (20%) (Fegert and others 2002). Remember that you must establish a good rapport with the audience in order for your oral presentation to be successful. The following guidelines will help.

Presentation Style

- Number your notecards so that if you drop them you can reassemble them quickly.
- Dress appropriately for the occasion.
- Use good posture. Good posture is equated with self-assurance, while slouching implies a lack of confidence.
- Be positive and enthusiastic about your subject.
- Try to maintain eye contact with some members of the audience. Eye contact is perceived as more personal, as though the speaker is talking with individual listeners.
- Avoid distracting gestures and mannerisms such as pacing, fidgeting with the pointer, jingling the change in your pocket, and adjusting your hair and clothes.
- Speak clearly, at a rate that is neither too fast nor too slow, and make sure your voice carries to the back of the room. Avoid "um's," "ah's," and other nervous sounds.

Integrating Visuals

Visuals are an integral part of an oral presentation. If you use **transparencies**, be sure to:

- Number transparencies in chronological order. If they fall down, you can reassemble them easily.

- Avoid pointing your finger on the transparency itself (your finger will be a blur on the projection screen). Instead lay a pen on the transparency as a pointer, or point directly to the screen.

If you use **PowerPoint** to prepare your presentation:

- Do not stand behind the computer monitor and read your presentation off the screen. This practice distances you from your audience.

No matter which method you choose, the following points apply:

- Do not display a visual until you are ready to discuss it.

- Stand next to the projection screen and point with your outstretched arm or pointer (stick) or use a laser pointer.

- Tell the audience exactly what to look for. Interpret statistics and numbers so that they are meaningful to your listeners. If you are describing Figure 8.2, for example, "Beta-galactosidase production increased dramatically in only 10 minutes after adding the lactose analog" is likely to make a bigger impression on the audience than "The absorbance increased from 0.12 to 0.71 in only 10 minutes after adding the lactose analog."

- Point to the specific part of the visual that illustrates what you are describing. Avoid waving the pointer in the general direction of the visual.

- Do not block the projected image with your body.

- Do not turn your back on the audience.

Interacting with the Audience

A unique benefit of communicating information verbally is that you can adjust your presentation based on the feedback you receive from your listeners. Pay attention to the reaction of individuals in the audience. Does their posture or facial expression suggest interest, boredom, or confusion? If you perceive that the audience is not following you, ask a question or two, and adjust your delivery according to their response. Your willingness to customize your presentation to your audience will enhance your reputation and lead to greater listener satisfaction.

To keep listeners focused:

- Establish common ground immediately. Ask a question or describe a general phenomenon with which your audience is familiar.

- Make sure your introduction is well organized and proceeds logically from the general to the specific without any sidetracks.

- Alert your audience when you plan to change topics. Practice these transitions when you rehearse.

- Summarize the main points from time to time. Tell your listeners what you want them to remember.

- Use examples to illustrate your hypotheses and conclusions.

- End your presentation with a definite statement such as "In conclusion..." Don't fade out weakly with phrases such as "Well, that's about it."

Group Presentations

If you are presenting with another person, your individual contributions should complement one another, not compete with each other. This requires good coordination and practice beforehand.

Fielding Listener Questions

Allow time for questions and discussion. Take questions as a compliment—your listeners were paying attention! Listen carefully to each question, repeat it so that everyone in the room can hear, and then give a brief, thoughtful response. If you don't understand the question, ask the listener to rephrase it. If you don't know the answer, say so.

Signal the end of the question session with "We have time for one more question" or something similar.

Feedback

It takes not only practice, but also coaching, to become an effective speaker. When you make an oral presentation to your class, your instructor may ask your classmates to evaluate your delivery, organization, and visual aids, as well as the thoroughness of your research using a form such as the "Evaluation Form for Oral Presentations" available at <http: www.sinauer.com/Knisely>.

When you are in the position of the listener, make the kinds of comments you would find helpful if you were the speaker. As you know, feedback is most likely to be appreciated when it is constructive, specific, and done in an atmosphere of mutual respect.

Word Processing Basics in Microsoft Word

If you have been writing papers in Microsoft Word since middle school, you can probably skip Sections 1–3 of this appendix. You may find Section 3.6, "Tracking Changes Made by Peer Reviewers," helpful during the revision process. If you want to save time inserting Greek letters and mathematical symbols, italics, and subscripts and superscripts, or if you want to replace long chemical names with a shorter keystroke combination, look over Section 4, which covers special tasks in scientific papers.

The screen drops in this appendix are for Microsoft Word 2003 on a PC, but the instructions apply to earlier versions as well unless otherwise noted. Path notation for reaching a particular command or option is shown relative to a command on the menu bar, toolbar, or task bar. For example, **File | Page setup | Margins | Landscape** means "Select the File command (on the menu bar), then click Page setup, then select the Margins tab, and choose Landscape." These general instructions apply to Word on Macintosh computers as well, although there are some differences, such as the use of the Option key in place of the Control key.

1. Good Housekeeping Habits

1.1 File Organization

You can expect to type at least one major paper and perhaps several minor writing assignments in every college course each semester. This amounts to a fair number of documents on your hard disk and removable disks (CDs, zip disks, and floppy disks). Good file organization is important for several reasons:

- When you are writing papers in different subjects at different times during the semester, you need to have a simple, logical system that will allow you to find the files you are working on. Nothing is more frustrating than to have a deadline and waste precious time searching for the wanted file.

- If your instructor informs you that he/she never received your paper, you will be expected to print out another copy. Excuses such as "I can't find the file" or "I accidentally erased it" are not viewed favorably.

- If you are asked to be a teaching assistant for an introductory course, it is helpful to be able to refer to your old lab reports when students ask you questions. The compactness and convenience of an electronic copy can't be beat.

When setting up a filing system on your computer, think of the computer as a large filing cabinet, with the drawers representing the computer's drives. Typically the hard drive is the **C:** drive, the floppy drive is **A:**, and, if present, the CD (compact disk) and zip drives are given other letter designations.

In a paper filing cabinet, the drawers contain folders, the folders may be further subdivided into subfolders, and the subfolders contain the actual documents. Word creates a "My Documents" folder on the hard drive as a possible location for storing your files. You can create subfolders within "My Documents," or you can create folders in addition to "My Documents."

One way to organize your files is by course number (Figure A1.1). Within each course, you could create subfolders for different assignments, such as lecture notes, homework, or lab reports. If you expect to have only a few assignments that require work on the computer, you may not even need to make subfolders.

There are two easy ways to create folders or subfolders. One way is to use Windows Explorer, and the other way is to make folders within Word.

Creating folders in Windows Explorer. For PCs, click the **Start** button on the taskbar in the lower left corner of the screen. Then select **All Programs | Accessories | Windows Explorer**. Use the arrow button to select the drive where you want to create a new folder (see Figure A1.1). Double-click the drive to display the existing folders. If you want to add a new folder on this level, select **File | New Folder** from the menu bar. Give the folder a unique, meaningful name that you will be able to recognize even years from now. To add a subfolder to an existing folder, first double-click the *folder,* and then select **File | New Folder** from the menu bar.

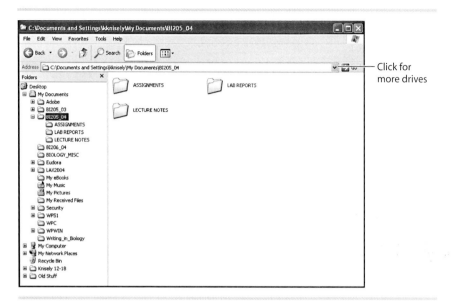

Click for
more drives

Figure A1.1 Possible ways to organize your files using folders and subfolders in Windows Explorer. In the Microsoft Office XP programs, the Tree Pane (overview of directories) on the left can be replaced by a Search pane if you wish to perform tasks on objects listed in the Details pane on the right. To do this, switch **View | Explorer Bar | Folders** to **View | Explorer Bar | Search**.

Creating folders in Word. When you open Word, a blank document appears. At the top of the screen, there is a Title bar, a Menu bar, and the Standard and Formatting toolbars (Figure A1.2). To open a previously created document, click **File | Open** and locate the file or choose a recently used file from the list.

To create a new folder in Word:

1. Click **File | Save As** on the menu bar. The **Save As** dialog box appears.

2. Click the down arrow next to the **Save in** text box to locate the folder in which you want to create a subfolder.

3. Click the **Create New Folder** button. The **New Folder** dialog box appears (Figure A1.3).

4. Type a name for the new folder and click **OK**.

The new folder now appears in the **Save As** dialog box.

(A)

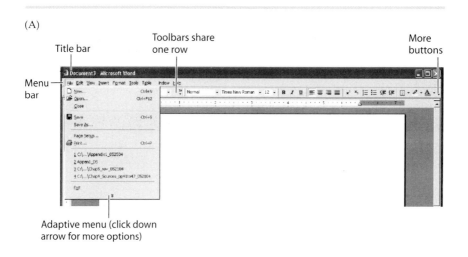

(B)

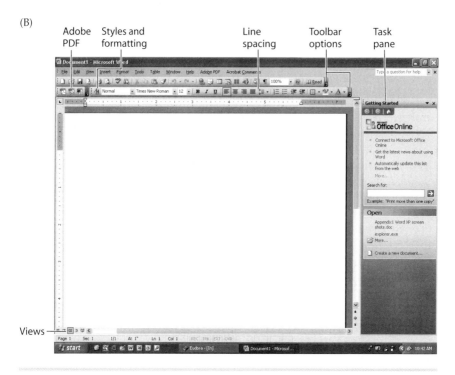

Figure A1.2 (A) Screen display of a blank Word 2000 document with adaptive menus and Standard and Formatting toolbars on one row. (B) Screen display of a blank Word 2003 document with the Task pane on the right. Some of the new features are labeled.

Create new folder

Save in text box

File extension (format)

Figure A1.3 Save As dialog box allows you to create folders and sub-folders in Word

1.2 Saving your Files

One of the most important tasks you must learn when working on the computer is saving your files. When you write the first draft of a paper by hand, you have tangible evidence that you have done the work. When you type something on the computer, however, your work is unprotected until you save it. That means that if the power goes off or the computer crashes before you save the file, you have to start again from scratch. Hopefully it won't take the loss of a night's work to convince you to make sure you know how to save your files!

Save your work often. The more difficult or complicated the text, the more often you should save it. Think about how long you would need to retype the text if it were lost and whether you can afford to spend that much time redoing it.

Naming and saving files manually. Although AutoSave (described in the next section) provides some insurance against lost work, saving your file manually produces a permanent file that will still be there if the power goes off or the computer crashes.

Do not wait until you are finished typing the document to save a file—save it after you have written the first sentence. Even though Word and other word processors save files in a temporary memory, your work is not protected unless it has been saved in a permanent form.

1. **To save your file for the first time,** click **File | Save As** on the menu bar (see Figure A1.2). The **Save As** dialog box appears (see Figure A1.3).

2. You have to make three important decisions when saving a file for the first time:
   - Location (drive, folder, and subfolder, if applicable)
   - File name
   - File extension (format)

The default **location** where the current document will be saved is shown in the **Save in** text box. To change to another folder or another drive, click the arrow next to the text box. Double-click the drive, folder, or subfolder until you arrive at the location where you want to save the current document. Only the folders for the selected drive will be displayed. To see the folders in another drive, double-click the desired drive.

You will need to choose a **file name**. The name suggested by Word is shown in the **File name** text box near the bottom of the dialog box. You may wish to replace Word's suggestion with a descriptive name that allows you to find it easily when you do future revisions. Sometimes you may also want to keep track of different versions of the same document by appending the file name with a number or letter (e.g., Protein1.doc, Protein2.doc, etc.).

Windows machines automatically assign a **file extension** (the three or four letters following the period in the file name) based on the program you are working in. It is best not to change the extension unless you expect to open the file in a different program. Some common file extensions are shown below. When you have entered the location, file name, and file extension, click **Save** to save the file and close the dialog box.

Program	Extension
Word	.doc
Excel	.xls
PowerPoint	.ppt
Adobe Acrobat	.pdf
Images	.jpg, .tif, .gif
HTML	.html, .htm, .htx

To save the same file subsequently, click the Save button (🖫) on the Standard toolbar, or **File | Save** on the menu bar. You can also save files using the keyboard by pressing Ctrl+S.

To save the same file at a different location, under a different name, or with a different file extension:

1. Click **File | Save As** on the menu bar (or press the F12 key). The **Save As** dialog box appears (see Figure A1.3).

2. Change the location, name, or extension as needed.

3. Click **Save** to save the file and close the dialog box.

AutoSave. **AutoSave** automatically saves the file you are currently working in at specified intervals. It should be considered a backup to (not a substitute for) saving the file manually.

1. To activate AutoSave, click **Tools | Options** on the menu bar (see Figure A1.2).

2. Select the **Save** tab in the **Options** dialog box (Figure A1.4). Select:
   - Allow background saves
   - Save AutoRecover info every: ___ Minutes (use the arrow to specify the interval, e.g., 10 minutes)
   - Save Word files as: Word Document (.doc)

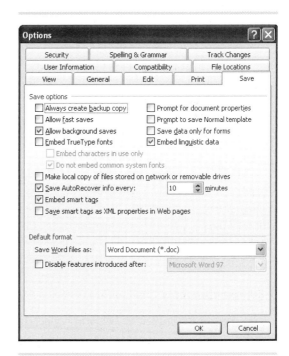

Figure **A1.4** Options dialog box, Save tab

The **Allow fast saves** option makes saving files faster by recording only those changes made in the document during the current session. The changes are then added (electronically) to the end of the file, instead of incorporating them into the file at the location where they were made. Fast saves increase the size of the document, so it is a good idea to turn off this option when you save the final version of a file. Disallowing fast saves forces Word to do a full save, in which the changes are incorporated properly, and the file size is reduced.

AutoSave saves the document in a **recovery file**. This feature is not a substitute for manually saving documents, because the recovery file is deleted when you save or close a document. If the power goes off, the recovery file still exists. When you restart Word, Word opens the recovery file so that you can save it manually. If you do not save the recovery file, it is deleted.

Backup files. It is very, very important to make at least one backup copy of your file while you are writing an important paper. The backup file should be saved on a removable disk, not on your hard drive. If your hard drive goes down, having a backup copy on a removable disk means you have not lost your work; you can continue working on your paper on another computer.

To make a backup file, follow the first set of instructions in Section 1.2. Change the location to one of your computer's removable disk drives or use a USB Flash Drive (also called JumpDrive or Thumb drive) to transfer files from one computer to another. Depending on capacity, flash drives hold as much as or more than a CD, yet they are only the size of a stick of gum.

1.3 Protecting your Files Against Viruses and Worms

Keep your antivirus software up to date by downloading the most recent software from the company's Internet site. New viruses and worms infect someone's computer every day, so you must aggressively protect your system and all the time invested in writing your papers. An ounce of prevention is worth a pound of cure, as the saying goes.

1.4 Removing Files

If you use your computer a lot, and especially if you have large files and graphics, the disk space fills up fast. Your computer will give you a message when you are running out of disk space. That is a sign that you need to remove files from the hard drive(s).

If you have already backed up an inactive file (one on which you are no longer working) on a removable disk, then you can delete this file from the hard drive. Also check for files in the Recycling Bin, as well as for temporary files, files attached to e-mail messages, and images saved from the Internet. Copying these files to removable disks and then deleting them from the hard disk frees up valuable space on your computer.

Deleting files using Word. You can easily delete individual files in Word using these steps:

1. Click **File | Open** on the menu bar.
2. Locate the file you want to delete (you may have to change drives or folders).
3. Right-click the unwanted file, and select **Delete**.

Removing or archiving files using Windows Explorer. If you need to remove a lot of files from your computer, use Windows Explorer. Click the **Start** button on the taskbar in the lower left corner of the screen and select **All Programs | Accessories | Windows Explorer**. Start with the **C:** drive, and look in each folder for inactive files.

1. **To delete files without copying them,** simply select the unwanted file by clicking on it. Then click **File | Delete** and "Yes" to confirm that you want to delete that file.
2. **To archive a file,** copy it to a removable disk as follows:
   a. Insert a CD, floppy, or zip disk into the respective drive.
   b. Click **Start | All Programs | Accessories | Windows Explorer** to open a second Explorer window.
   c. Use the arrow button to select the drive you want to copy to.
   d. Double-click the drive to display the contents of the disk.
   e. Locate the file(s) you want to copy on the hard drive. To copy a single file, select it by clicking the file name. To select several consecutive files, click the first one, hold down the **Shift** key, and click the last one. To select several non-consecutive files, click each file while holding down the **Ctrl** key.
   f. Click **Edit | Copy** from the menu bar. Alternatively, right-click the mouse and select **Copy**.
   g. Click the Explorer window displaying the contents of the removable disk.
   h. Click **Edit | Paste** from the menu bar. Alternatively, right-click the mouse and select **Paste**. The copied file will be displayed in the window for the removable disk drive.

An even faster way to copy is to select the file(s) to be copied and drag it (them) to the new location. **This method replaces steps f through h.**

2. Formatting a Document

All scientific journals have a section devoted to instructions for authors. These instructions tell the author exactly how to format a paper, so that all the papers in the journal have a uniform appearance. If your instructor has not given you specific instructions, the layout in Table A1.1 will lend a professional look to your lab report or scientific paper. If you are already familiar with Word, the Quick Reference tells you where to make these formatting adjustments. Detailed instructions are given in the sections following the table.

2.1 Menu Bar and Toolbars

When you open Word, a blank document appears. At the top of the screen, there is a Title bar, a Menu bar, and one or two toolbars. To see the hidden buttons, click More Buttons (see Figure A1.2A) or Toolbar Options (see Figure A1.2B).

Word 2000 and later versions have so-called adaptive menus that show only the basic commands (as decided by Word); the rest of the commands are displayed after five seconds, or you can click the down arrow to see them without waiting. If you are accustomed to having all of the buttons and menu commands available, you will find toolbars on one row and adaptive menus a nuisance.

To personalize your menus and toolbars, click **Tools | Customize | Options**, and clear or select the check boxes according to your preferences.

Toolbars consist of pictorial buttons that are shortcuts to menu commands. If you place the mouse pointer over a particular button (do not click any mouse buttons), a description of that button will appear. For example, in Word 2003 if you place the mouse pointer on the fifth button from the left on the Standard toolbar (🖨), the Print command, with the default printer in parentheses, will appear. The Print command is also found on the menu bar under **File | Print**.

Note: Shortcuts offer only one option from the corresponding menu command. For example, if you click the **Print** button on the standard toolbar, the entire document will be printed on the default printer. If you want to print single pages or use a different printer, you must go to the menu bar and select the desired options from the **File | Print** pull-down menu.

The menu bar commands that are important for formatting your document are **File**, **Insert**, and **Format** (Figure A1.6).

TABLE A1.1 Giving your document a professional appearance

FEATURE	LAYOUT	QUICK REFERENCE
Paper	8½ x 11" (or DIN A4) white bond, one side only	File | Page setup | Paper | Letter
Margins	1.25" left and right; 1" top and bottom	File | Page setup | Margins
Font size	12 pt (points to the inch)	Formatting toolbar | Font size
Typeface	Times Roman (serif) or Helvetica (sans serif). Serif fonts are easier to read for most people, but sans serif fonts are favored in graphics.	Formatting toolbar | Font See also Format | Style
Pagination	Arabic number, top right on each page except the first	Insert | Page numbers
Page Break	To maintain section continuity	Ctrl+Enter or Insert | Break | Page Break
Spacing	Double	Formatting toolbar | Line spacing
New paragraph	Indent 0.5"	Format | Tabs | Default
Aligning text	**Justification** is Left/ragged right or Justify/even edges	Format | Align left or Justify
	Indent entire paragraph to set off from rest of text	Format | Paragraph | Left and right indentation
	Tabs	Format | Tabs
	Table: Create a table	Table | Insert table
	Lists that begin with **Bullets** or **Numbers**	Format | Bullets and Numbering
Title page (optional)	Title, authors (your name first, lab partner second), class, and date	
Headings	Align headings for Abstract, Introduction, Materials and Methods, Results, Discussion, and References on left margin or center them. Use consistent format for capitalization.	Format | Style | Heading 1.
Subheadings	Use sparingly and maintain consistent format	Format | Style | Heading 2, Heading 3, etc.
References	Citation-Sequence System: Make a numbered list.	Format | Bullets & Numbering | Numbered tab
	Name-Year System: Use hanging indent to list references.	Format | Paragraph (Indents and Spacing tab) | Special | Hanging by 0.25"
	Both systems: Use accepted punctuation and format.	
Assembly	Pages in order, staple top left corner	

Word 2003 Update

The Microsoft Office XP programs Word 2002 and Word 2003 look different from their predecessors – the new default window is split into the Document pane and the Task pane (Figure A1.5).

The Task pane appears when you choose certain commands or when you click **View | Task Pane** on the menu bar. To switch to another Task pane, click the Title bar at the top of the Task pane. To close the Task pane, click the Close box (☒) in the top right corner of the Task pane.

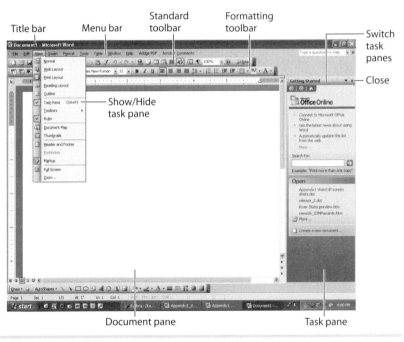

Figure A1.5 Screen display of a blank Word 2003 document

2.2 Paper

Click **File | Page Setup** on the menu bar. You will find three tabs: **Margins, Paper**, and **Layout** (Figure A1.7). In Word 2003, you choose the orientation of the paper (portrait vs. landscape) on the **Margins** tab. In Word 2000, the orientation is selected on the **Paper Size** tab. To

Figure A1.6 Menu bar with arrows on commands used to format a Word document

change paper size, click the **Paper** tab, then click the arrow button and make your selection.

2.3 Margins

Click **File | Page Setup** on the menu bar (see Figure A1.7). On the Margins tab, the default settings for the top and bottom margins are 1" and those for the left and right margins are 1.25". The Gutter refers to the amount of space between multiple columns on a page.

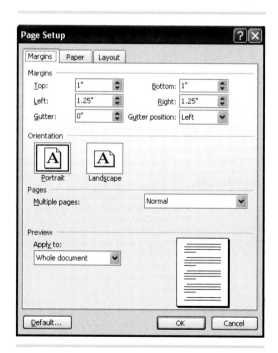

Figure A1.7 Page Setup dialog box. The Paper Size and Paper Source tabs (Word 2000 and earlier versions) are combined on the Paper tab in Word 2003.

2.4 Font Style and Point Size

Serif fonts such as Times Roman are easier to read for most people than sans serif fonts such as Arial. The most legible point size generally seems to be 12. Thus, for most machines, the default Normal font is 12 pt Times New Roman.

To modify the font style, size, or appearance, click **Format | Font** on the menu bar or one of the buttons on the Formatting toolbar (Figure A1.8). To apply formatting to existing text, you must first select the text that you want to format. To select a word, double-click it. To select a paragraph, triple-click anywhere within the paragraph. To select a block of text, use one of the methods described in Section 2.7.

2.5 Page Numbers

Click **Insert | Page Numbers** on the menu bar to open the dialog box (Figure A1.9).

In the **Page Numbers** dialog box, select the position (top or bottom of the page) and the alignment (left, center, right, inside, or outside) of the page numbers. Journal articles tend to have the page numbers placed at the bottom, on the outside margins. For your course work, however, top right is fine. Be sure to clear the **Show number on first page** check box, as it is not customary to put a number on the first page. All pages of your paper except the first page should have a page number, so that you can easily determine the order of the pages and whether all pages are present.

By convention, Arabic numerals are used to number the pages. This style is the default page number format in Word. In some special situations (e.g., Table of Contents or Preface of a book), Roman numerals or letters are used. These alternative formats can be selected by clicking the **Format** button.

The font of the page number should be the same as the font of your text. Because page numbers are part of the header or footer, you can check the page number font by displaying the header and footer. Click **View | Header and Footer**, or click **View | Print Layout.**

Figure A1.8 Formatting toolbar. Adobe PDF buttons have been added to the left side of this toolbar in Word 2003.

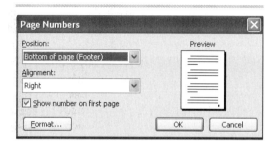

Figure A1.9 Page Numbers dialog box

Double-click the page number to view the current font, font style, or font size. Modify the format by using the corresponding options on the Formatting toolbar (see Figure A1.8).

To modify the page number position or alignment at any time, click **Insert | Page Numbers** on the menu bar.

2.6 Page Breaks

A **page break** starts the text following the break on the next page. For example, let's say that you start a new section of your lab report or scientific paper near the bottom of a page. Perhaps the section heading fits, but the rest of the text gets thrown onto the next page. To make sure that the section heading is not separated from the rest of the text, position the cursor just in front of the heading. Click **Insert | Break | Page break**. A faster way to do this is to put the cursor just in front of the heading and press Ctrl+Enter on the keyboard.

2.7 Line Spacing

Line spacing refers to the amount of space between lines on a page and is found under **Format | Paragraph** (**Indents and Spacing** tab) on the menu bar. The Formatting toolbar in Word 2003 also has a **Line Spacing** button (see Figure A1.2B). The default setting is Single. To change line spacing before you begin typing your paper, select Double under **Line Spacing** in the **Paragraph** dialog box (Figure A1.10) or click the **Line Spacing** button and select "2." To adjust line spacing at a later time, **select the block of text** to be changed by either of the following methods:

- Hold down the left mouse button, and drag the cursor across the text from the first character to the last.

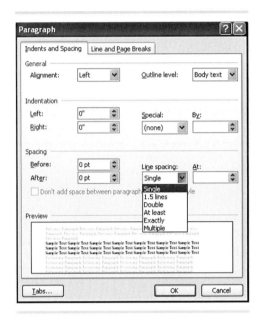

Figure A1.10 Paragraph dialog box, Indents and Spacing tab

- Click the left mouse button at the beginning of the text, hold down the **Shift** key, and click the last word in the block. This method works well when the text extends over several paragraphs or even pages.

When the block of text is selected, follow the aforementioned procedure to adjust line spacing.

2.8 Aligning Text

There are a number of different ways to align text in a Word document. The ones you are most likely to use are as follows:

- Justification (right, left, center, or full). **See Section 2.9.**
- Indenting an entire paragraph so that it is set off from the rest of the text. **See Section 2.10.**
- Tabs. **See Section 2.11.**
- Table columns (creating a table). **See Section 2.12.**
- Lists that begin with numbers or bullets. **See Section 2.13.**

Never use the space bar to align text, because variation in character size will lead to variation in vertical text alignment as well.

2.9 Justification

Justification is the adjustment of the lines so that they are aligned either on the left margin (with ragged right edge), centered between the margins, aligned on the right margin (with ragged left edge), or aligned both on the left and right margin (with even edges). Under **Format | Paragraph**, on the **Indents and Spacing** tab, select either **Left** or **Justify** next to **Alignment** (see Figure A1.10). These options are also available on the Formatting toolbar (see Figure A1.8).

2.10 Indentation

Indenting a block of copy allows you to change the margins, so that only that block of copy will be set off from the others.

To indent copy, click **Format | Paragraph** on the menu bar. On the **Indents and Spacing** tab (see Figure A1.10), set the left and right Indentation as desired (for example, 1"). When you type the text, it will be set off from the rest of the text with wider margins. Pressing **Enter** returns the indentation to the default values.

If you have already typed a passage that you later decide you want to indent:

1. Select the passage by one of the methods described in Section 2.7.
2. Click **Format | Paragraph**.
3. Set the desired indentation on the **Indents and Spacing** tab.
4. Click **OK**.

2.11 Tabs

Tab stops allow you to jump a set distance forward on a line. It is always preferable to use tab stops rather than the space bar to align text in columns. Because character spacing varies with the size of the characters in a font, different words, even though they have the same number of characters, occupy different-sized spaces on a line. Adding more spaces will not change this mismatch. Instead, put the words on a tab stop, which defines the vertical alignment regardless of the characters. Remember, **never use the space bar to align text in columns!** (For multiple columns, also see **Section 2.12,** "Tables with no lines.")

To adjust tab stops, select **Format | Tabs** on the menu bar. The default tab stops are given in the top right corner of the **Tabs** dialog box and will most likely be 0.5". Change the default if desired, or add individual tab stops as needed.

The indentation that designates a new paragraph is determined by the first tab stop in the default setting. The customary paragraph indentation is 0.5″.

2.12 Creating a Table

By convention, tables in scientific papers do not have vertical lines to separate the columns, and horizontal lines are used only to separate the table caption from the column headings, the headings from the data, and the data from any footnotes (see, for example, Table 1 in Figure 4.2).

To create a table:

1. Position the cursor where you want to insert the table.
2. Click **Table | Insert Table** on the menu bar (Figure A1.11). Enter the number of columns and rows.
3. To change the table format, click anywhere in the table and then select **Table | Table AutoFormat**. "Simple 1" is an appropriate table format for a scientific paper.
4. Click **OK**.

A blank table appears with the cursor in the first cell.

Tables with no lines. To align text in columns without showing any lines:

1. Click **Table | Insert Table** on the menu bar. Enter the desired number of columns and rows.
2. Under **Table | Table Properties** on the **Table** tab, click the **Borders and Shading** button.

Figure A1.11 Table commands

3. On the **Borders** tab, select **None**.

4. Click **OK**.

Navigation in tables. To **jump** from one cell to an adjacent one, use the arrow keys. To move forward across the row, use the **Tab** key. *Note:* If you press **Tab** when you are in the last cell of the table, Word adds another row to the table.

To align text on a tab stop in a table cell, press Ctrl+Tab.

Inserting and deleting rows or columns. To insert a new column or row, position the cursor in a cell adjacent to where you want to insert a new column or row. Select **Table | Insert** and then the desired location (Columns to the left, Columns to the right, Rows above, or Rows below).

To delete a column, row, or individual cell, position the cursor in the column, row, or cell you want to delete. Select **Table | Delete** and make your selection (see Figure A1.11).

Changing column width or row height. To change the width of a column, position the cursor on one of the vertical lines so that ◂╫▸ appears. Then hold down the left mouse button and move the line to make the column the desired width.

To change the height of a row, position the cursor on one of the horizontal lines so that ╪ appears. Then hold down the left mouse button and move the line to make the row the desired height.

Other table commands. Position the cursor in the table, and then click **Table** on the menu bar. Other commands for customizing your table are shown in the pulldown menu (see Figure A1.11).

You can format the text in each cell just as you would format running text. For example, to center the column headings or to put them in boldface, select the top row and click the **Center** or **Bold** button on the Formatting toolbar. To align numbers, select the entire column and click the **Align Right** button. To format all the cells, click the top left cell, hold down the **Shift** key, and click the bottom right cell. To format selected cells, do the same, except hold down the **Ctrl** key.

2.13 Lists That Begin with Numbers or Bullets

A **numbered list** may be used when you want to list references in the Citation-Sequence system or steps in a procedure or organize your thoughts first as an outline. A **bulleted list** is handy for highlighting or summarizing the main points of a paper or poster, when chronological order is not important.

To make a bulleted list:

1. Click **Format | Bullets and Numbering** on the menu bar. You will see four tabs in the **Bullets and Numbering** dialog box: Bulleted, Numbered, Outline Numbered, and List Styles (Figure A1.12).

2. Choose the **Bulleted** tab to make an unnumbered list. You have a choice of eight bullets: none and symbols such as filled dots, open dots, arrows, and boxes. Click the one you want.

3. If you prefer a different bullet, click any box except none, and then press the **Customize** button at the bottom right of the dialog box. In the **Customize Bulleted List** dialog box (Figure A1.13), click the **Character** button. This opens the **Symbols** dialog box, which allows you to change the font to any ASCII character.

4. Again in the **Customize Bulleted List** dialog box, click the **Font** button. This allows you to change the size or style of the bullet.

5. Finally, set the indentation of the bullet and the text. If you want the bullet to start on the left margin, set the indentation to 0″. Set the text indentation 0.25″ more than the bullet indentation. Click **OK**.

6. When you have finished typing the list, press **Enter**, and then **Backspace** to delete the bullet on that line.

You can also type a list first, and then add bullets later on. To do this, select the passage by one of the methods described in Section 2.7, and then select **Format | Bullets and Numbering**, and the desired format.

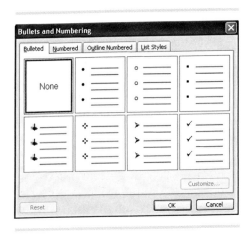

Figure A1.12 Bullets and Numbering dialog box, Bulleted tab

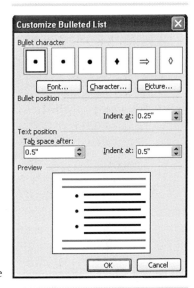

Figure A1.13 Customize Bulleted list dialog box

If you want to make a numbered list:

1. Click **Format | Bullets and Numbering** on the menu bar. You will see four tabs in the Bullets and Numbering dialog box: Bulleted, Numbered, Outline Numbered, and List Styles (see Figure A1.12).

2. Choose the **Numbered** tab (Figure A1.14). You have a choice of eight options: none and different styles of numbers or letters. Click the one you want.

3. If you prefer another option, click any box except none, and then press the **Customize** button at the bottom right of the dialog box. In the **Customize Numbered List** dialog box, set the Number format, style, and position, and the Text position (0.25″ more than the number position).

4. Back in the **Bullets and Numbering** dialog box (**Numbered** tab), select how you want the list numbered: **Restart numbering**, or **Continue previous list**. If you didn't have any previous numbered list of that format, these selections will not be available.

Even though it is practical to number the steps in a procedure for your own reference, this is never done in scientific papers. The procedures are always written in paragraph form in the Materials and Methods section, never as a numbered list.

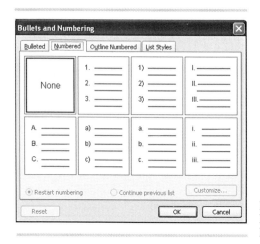

Figure A1.14 Bullets and Numbering dialog box, Numbered tab

2.14 Style

A **style** is a set of formats that can be applied to a few words, a paragraph, or a whole document. Look on the left side of the Formatting toolbar for a box containing the word *Normal* (see Figure A1.8).

Click the down arrow next to the word Normal to see what font typeface and size are currently specified for the normal text (running text of the document). In this pulldown menu, you can also see what styles are specified for different levels of section headings and for the header and footer.

Scientific papers and laboratory reports are divided into sections, and each section begins with a heading (for example, **Introduction**). To ensure that your section headings are uniform, click the down arrow next to the Style box and select Heading 1. Formatting is automatically applied as you type the heading. Press **Enter** and the formatting of the subsequent text returns to normal. For subheadings, select Heading 2, Heading 3, and so on.

Headings are also used to generate a table of contents, so you might find this feature useful for longer research papers.

To change the characteristics of a style:

1. Click the **Styles and Formatting** button on the left of the Formatting toolbar to open the Task pane (Figure A1.15).
2. On the **Task** pane, right-click the style you want to modify.
3. Click **Modify** to open the **Modify Style** dialog box and change the characteristics as needed.
4. Click **OK**.

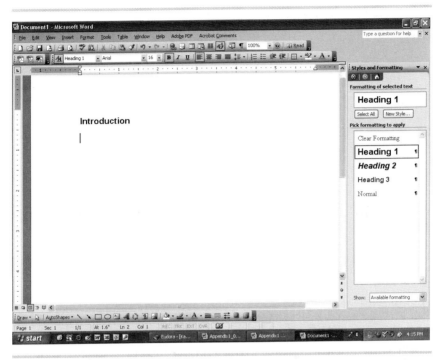

Figure A1.15 Style dialog box with Styles and Formatting task pane displayed

In Word 2000, these changes are made by selecting **Format | Style**.

2.15 Templates

Templates specify the format of all Word documents. When you select **File | New** from the menu bar or the **New Blank Document** button on the Standard toolbar, Word opens a new document based on the **Blank Document** template. This is the default template.

 If your instructor has specified a particular format for all your papers, you can customize an existing template for the format you need. Follow these steps:

1. Select **File | New** from the menu bar.

2. In Word 2003, the Task pane displays the **New Document** options. Click **On my computer** under **Templates** to display the templates available on your computer. In Word 2000, the Templates window is opened automatically (Figure A1.16).

3. Click the blank document icon on the **General** tab.

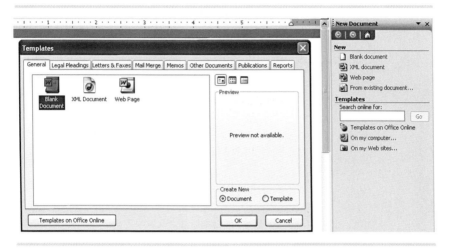

Figure A1.16 Templates available under the General tab. In Word 2003, this window is opened by clicking "On my computer" on the Task pane.

4. Click the **Template** option button under **Create New**.
5. Check the settings for each feature in Table A1.1, or your instructor's specifications, and make changes where necessary.
6. Select **File | Save As** from the menu bar. The **Save As** type will automatically be **Document Template (*.dot)**.
7. Type the name of the new template next to File name. For example, you might give the template the same name as your course.

Every time you type papers for that course, select **File | New**, and then click the course icon. Using the same template gives your documents a uniform appearance.

3. Editing the Text

You will modify parts of your document while you are writing it, as well as afterwards during the editing and proofreading stages. Some of the most common modifications involve:

- Views
- Cut, copy, and paste
- Find and replace
- Spelling and grammar checker

TABLE A1.2 Quick reference for editing commands

FEATURE	QUICK REFERENCE
Views	**Word 2000:** View | Normal / Web Layout / Print Layout / Outline / Toolbars / Ruler / Document Map / Header and Footer / Full Screen / Zoom **Word 2003:** Reading Layout / Task Pane / Thumbnails / Markup are new
Zoom	Standard toolbar | Select percentage or another option
Cut, Copy, and Paste	Standard toolbar | ✂ , 🖹 , 📋 Keyboard Shortcuts: Ctrl + X (cut), Ctrl + C (copy), Ctrl + V (paste)
Find, Replace	Edit | Find or Replace Keyboard Shortcuts: Ctrl + F (find), Ctrl + H (replace),
Automatic spelling and grammar checker	Tools | Options | Spelling and Grammar
Manual, systematic spelling and grammar checker	Standard toolbar | ABC✓ button
Word count	Tools | Word Count
Tracking changes made by peer reviewers	**Word 2000:** Tools | Track Changes | Highlight Changes | Track changes while editing **Word 2003:** Tools | Track Changes automatically turns tracking on and off

- Word count
- Tracking changes made by peer reviewers

Use Table A1.2 if you are already familiar with these features. Detailed instructions follow the table.

3.1 Views

When you click **View** on the menu bar, you are given a number of choices that affect how your document appears on the screen (Figure A1.17). **Normal** is useful when you are interested in seeing only the text, but not the headers, footers, blank areas, or figures (images or graphs imported into the document). If you want a more realistic view of what the page

(A)   (B)

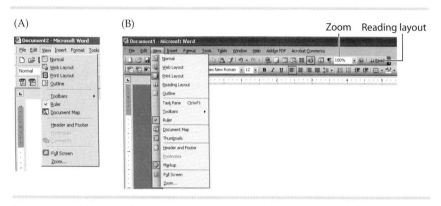

Figure A1.17 Comparison of View menus in (A) Word 2000 and (B) Word 2003

will look like with headers, footers, blank areas, and figures, then use the **Print Layout** view. To see two pages side-by-side, select **View | Reading Layout**.

To see which toolbars are currently displayed, click **Toolbars**. Those with a check mark next to them are displayed.

To edit the header and footer, click **Header and Footer**.

To adjust document size on the screen, click **Zoom**. You can type in a percentage, or select an option from the list. Choose a size that makes it easy for you to read the characters on the screen. The Zoom box is also located on the Standard toolbar, at the top right of the screen (see Figure A1.17).

3.2 Cut, Copy, and Paste

To use the cut and copy functions, you must first select the text that you want to cut or copy. To select a word, double-click it. To select a paragraph, triple-click anywhere within the paragraph. To select a block of text, use one of the methods described in Section 2.7. Then click either the **Cut** button ✂ or the **Copy** button 🗎 on the Standard toolbar or use the keyboard shortcuts Ctrl+X (cut) or Ctrl+C (copy). Move the cursor to where you want to paste this text. It can be pasted into the same document or into another document. Click the **Paste** button 🖺 or use Ctrl+V.

3.3 Find and Replace

Sometimes you realize that you have been using the wrong word or wrong symbol, or consistently misspelling a word, throughout a paper. This error can be easily remedied using the **Replace** option. Click **Edit**

| **Replace** on the menu bar. Type in the wrong word next to **Find what**, and the correction next to **Replace with**. If you are *absolutely certain* that you want the change to be made *every time* it appears in the document, then select **Replace all**. If the "wrong word" may not be wrong every time, select **Find next**. This allows you to preview the sentence in which the word was used and gives you the option *not* to replace it in every instance.

When you write the first draft of a paper, you may want to get all your ideas out first, without worrying about details or accuracy. It is important in subsequent drafts, however, to get the facts straight. You should devise your own way of marking the text that needs special attention. For example, you might type **??** after such text. Then when you want to go back and revise these sentences, you simply click **Edit | Find**, and type in **??**.

3.4 Spelling and Grammar

There is absolutely no excuse for typos when Word offers you so many options for checking spelling and grammar:

- AutoCorrect
- Automatic spelling and grammar checker
- Manual, systematic spelling and grammar check

AutoCorrect. The **AutoCorrect** function corrects common types of spelling mistakes as you type. To see the list of commonly misspelled words that Word corrects automatically, click **Tools | AutoCorrect**, and scroll through the words on the **AutoCorrect** tab. If you do not want AutoCorrect to change a particular keystroke combination (e.g., do not change (c) to ©), select this combination on the list and click the **Delete** button. You can also add words that you know you misspell frequently to the list.

If AutoCorrect changes text that you don't want changed while you are typing, click **Edit | Undo AutoCorrect** on the menu bar.

Automatic spelling and grammar checker. The automatic spelling and grammar checker underlines in red the words that are not in Word's spelling dictionary, and underlines in green possible grammatical errors.

To deal with a word underlined in red, click it with the *right* mouse button. A pop-up menu appears with commands and suggestions for replacements (Figure A1.18). Select **Add** after you have confirmed the correct spelling in your textbook or laboratory manual. The red underline

Insert a cuvette into the spectrophotometer.

Figure A1.18 Dialog box for spelling suggestions and other ways to handle words not in Word's dictionary

is deleted, and the word is ignored in the manual, systematic spelling and grammar check.

To deal with a possible grammatical error underlined in green, click it with the *right* mouse button. A pop-up menu appears (Figure A1.19).

To tell Word how aggressively to check spelling and grammar as you type, click **Tools | Options** on the menu bar. In the **Options** dialog box (Figure A1.20), choose the **Spelling & Grammar** tab.

Check the boxes of the options you want. For example, under **Spelling** you might check:

- Check spelling as you type
- Always suggest corrections

The egg white sample were serially diluted.

| sample was |
| samples were |
| Ignore Once |
| Grammar... |
| About This Sentence |
| Look Up... |
| Cut |
| Copy |
| Paste |

Figure A1.19 Dialog box for editing possible grammatical errors

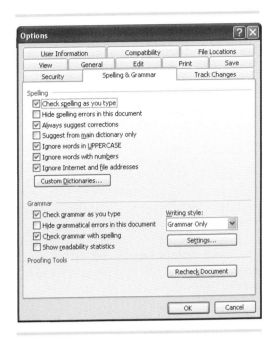

Figure A1.20 Options dialog box, Spelling & Grammar tab

- Ignore words in uppercase
- Ignore words with numbers
- Ignore Internet and file addresses

Under Grammar, you might select the following options:

- Check grammar as you type
- Check grammar with spelling

When you click the **Settings** box, you have the opportunity to select various grammar and style options.

Manual, systematic spelling and grammar check. To do a manual, systematic spelling and grammar check at any time, click **Tools | Spelling and Grammar**. The checker begins at the position of the cursor, and stops at every location at which it doesn't recognize a word or sentence construction. You then have the option of changing the word/construction, ignoring it, or adding the word to Word's dictionary. When the spell checker stops at a scientific word, consult your textbook for the correct spelling, as specialized terminology is not included in Word's dictionary.

As good as these features are, remember that **a spell checker is no substitute for a human proofreader**. Spell checkers cannot recognize mistakes of misuse, such as *there* and *their,* and they are not familiar with expressions such as *pH*. In fact, the spell checker may even try to get you to change a correctly used scientific word to a word that is in its repertoire, but that obviously makes no sense. Be vigilant and use common sense!

3.5 Word Count

The length of your lab report or paper depends on the detail expected by your instructor. Most instructors prefer not to specify a page length, stating simply that the work should be complete and concise. As a general rule, shorter is better.

If you do honors or graduate research later on, you may be expected to submit a proposal of a certain length. If the specifications are given in terms of words, the word count feature is helpful. Click **Tools | Word Count**, and Word will tally the number of words in the document.

3.6 Tracking Changes Made By Peer Reviewers

It may not always be possible for you and your peer reviewer (classmate) to find a common time to go over your paper. E-mail makes the peer review process more convenient. You can send your paper to your peer reviewer in electronic format as an attached file, your peer reviewer can make comments directly in your file, and then the reviewer can send it back to you.

It is important for you to be able to distinguish your original text from the comments and changes suggested by your peer reviewer. After all, *you* are the author, and you have the right to accept or decline a reviewer's suggestions.

The following instructions apply to Word 2000; see the Word 2003 Update box (on pp. 160–161) if you have a more recent version of Word. Before you send your lab report to a peer reviewer, click **Tools | Track Changes** and select **Track changes while editing** (Figure A1.21). As long as this box is selected, anything someone types in the document will be underlined and be in a colored font. Anything someone deletes will be crossed out and be in a colored font.

When you get your peer-reviewed document back, open it and select **Tools | Track Changes** on the menu bar, and then **Accept or Reject Changes**. The **Find** arrow will take you to the locations where changes were made, and then you can accept or reject them. Alternatively, go to **View | Toolbars** and display the Reviewing toolbar by clicking the check box. The commands on the toolbar are nearly the same as those in

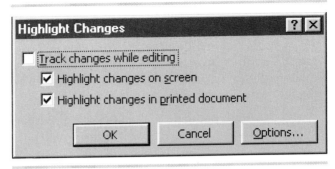

Figure A1.21 Highlight Changes dialog box in Word 2000

the **Accept or Reject Changes** dialog box (**Accept All** and **Reject All** are missing).

To see which reviewer made changes in your text, hold the cursor over the change for a second to display the ScreenTip showing the author, date and time, and type of change. If you are reviewing some-one else's paper, you can identify yourself by entering your name and initials in **Tools | Options | User Information**.

Make sure to print out the revised version of your paper, and proof-read the hard copy. Some mistakes are more easily identified on paper, and it is always better for you, rather than your instructor, to find them.

4. Special Functions in Scientific Papers

Scientific papers possess features that are usually not found in papers written in the humanities. These include:

- Greek letters and mathematical symbols
- Superscripts and subscripts
- Italics to indicate the scientific name of an organism
- Tables inserted in the document
- Figures imported from another program such as Excel
- Equations

In addition, you may need to leave space in order to add hand-drawn sketches to your paper. Another spacing requirement is hanging indents, which are used to separate references in name-year format. You can use the Quick Reference Table (Table A1.3) if you are already familiar with these features.

Word 2003 Update

Select **Tools | Track Changes** to turn this option on and off. When **Track Changes** is on, the Reviewing toolbar is displayed under the other toolbars (Figure A1.22). The Reviewing toolbar has commands that allow you to:

- Select how comments and changes (markup) are displayed

- Review each change in sequence

- Accept or reject each change or all changes

- Insert comments

- View or hide the Reviewing pane

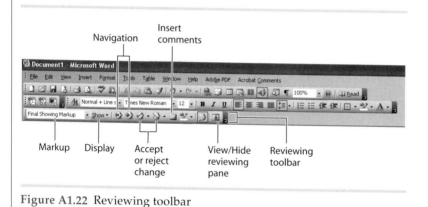

Figure A1.22 Reviewing toolbar

4.1 Greek Letters and Mathematical Symbols

Some instructors may allow you to write Greek letters and mathematical symbols in by hand. The problem with this approach is that it is easy to miss one of the places where these characters should be inserted in your text, and this oversight may cost you points on your grade. Furthermore, handwritten characters make the paper look unprofessional.

Take some time to locate the symbols you need before you write your paper.

1. Click **Insert | Symbol** on the menu bar.

Word 2003 Update (*continued*)

In previous versions of Word, markup was shown directly in the text, but now you have a number of options (Figure A1.23):

- Markup directly in the text (**Show | Balloons | Never**)
- Balloons along the right margin (**Show | Balloons | Always**)
- Separate Reviewing pane below the Document pane (**Show | Reviewing Pane**).

When you get your peer-reviewed document back, open it and make sure **View | Markup** is turned on. If the Reviewing toolbar is not displayed, click **View | Toolbars | Reviewing**.

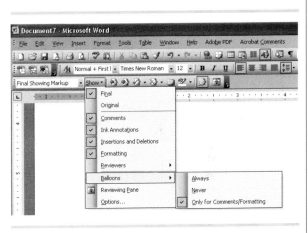

Figure A1.23 Options for displaying markup

2. Select **Font: (normal text)** and browse the available characters in this character set (Figure A1.24). You will find commonly used scientific symbols such as:

Symbol	Meaning
°	degree(s)
±	plus or minus
μ	micron (10^{-6} m)
‰	parts per thousand
f	function (calculus)

TABLE A1.3 Quick reference for features used in scientific papers

FEATURE	QUICK REFERENCE
Greek letters	Insert | Symbol | Font → (normal text) or Symbol
Mathematical symbols	Insert | Symbol | Font → (normal text) or Symbol
Superscripts and subscripts	Format | Font → Superscript or Subscript Keyboard shortcuts: Ctrl + (Subscript); Ctrl + Shift + (Superscript)
Italics	I button on Formatting toolbar Keyboard shortcut: Ctrl + I
Customize toolbar	Tools | Customize | Commands tab | Categories: Format | Scroll to desired command | Click command and drag it to desired location on Formatting toolbar
Shortcut keys	Insert | Symbol | Font → (normal text) or Symbol | Select character | Shortcut Key | Define combination of keystrokes | Assign
AutoCorrect	Tools | AutoCorrect | Replace ~ With ~
Tables inserted in the document	Table | Insert Table
Figures imported from another program such as Excel	See Appendix 2
Equations	Insert | Object | Microsoft Equation 3.0
Measuring space	Word processor's ruler
Hanging indent to list references	Format | Paragraph (Indents and Spacing tab) | Indentation: Special | Hanging

In the Normal text character set, you will also find characters with diacritical marks used in European languages other than English. Some common letters with diacritical marks are:

Name	Example
Accent	À, Á, È, É, Ì, Í, Ò, Ó, Ù, Ú, à, á, è, é, ò, ó, ì, í, ù, ú
Circumflex	Â, Ê, Î, Ô, Û, â, ê, î, ô, û
Cedilla	Ç, ç
Tilde	Ã, Ñ, Õ, ã, ñ, õ
Umlaut	Ä, Ë, Ï, Ö, Ü, Ÿ, ä, ë, ï, ö, ü, ÿ

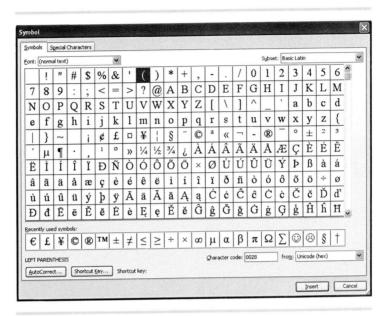

Figure A1.24 Symbol dialog box showing characters under Font: (normal text)

When you change the Font to Symbol (Figure A1.25), you will find upper- and lowercase Greek letters and other symbols used in mathematics and chemistry such as:

Symbol	Meaning
$\leq, \geq$	is less than or equal to, is greater than or equal to
∞	infinity
•	raised period (centered dot); multiplication symbol in equations; connection for adducts in a chemical formula
$\rightarrow, \leftarrow, \leftrightarrow$	arrows used to write chemical reactions
∂	partial derivative (calculus)

3. Once you have found the letter or symbol you need, simply click it and select the **Insert** button at the bottom of the dialog box.

If you use a Greek letter or mathematical symbol frequently, you can make a shortcut key to save time (see Section 4.5).

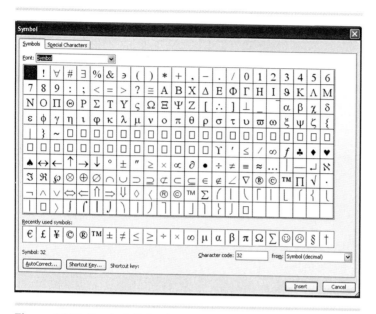

Figure A1.25 Symbol dialog box showing characters under Font: Symbol

4.2 Superscripts and Subscripts

Some expressions used in the natural sciences have superscripted or subscripted characters. It is *not correct* to write these characters on the same line as the rest of the text. Similarly, when using scientific notation, exponents are always superscripted. It is *not acceptable* to write exponents preceded by a caret (^) or by an uppercase E to designate superscript; these conventions have evolved for use in documents such as electronic mail in which formatting is not available. Do not use them in a printed paper.

    RIGHT: 2×10^{-3} (exponent is superscripted)

    WRONG: $2 \times 10{-3}$ or $2 \times 10\text{^}{-3}$ or $2 \times 10E{-3}$

To superscript or subscript text:

1. Type the expression without superscripting or subscripting.

2. Select the character(s) to be sub- or superscripted.

3. Click **Format | Font** on the menu bar (Figure A1.26).

4. Check the appropriate box under **Effects** in the **Font** dialog box.

5. Click **OK**. The text you selected is formatted.

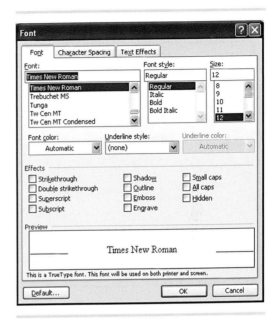

Figure A1.26 Font dialog box

If you use sub- and superscripts frequently, you can add buttons for these commands to the Formatting toolbar (see Section 4.4).

Alternatively, you can apply sub- and superscripting using keyboard commands. Select the character(s), then press Ctrl+ for subscript, or Ctrl+Shift+ for superscript.

4.3 Italics

By convention, scientific names given to organisms (viruses, bacteria, plants, and animals) are in Latin and are written in italics. After you type the name of the organism, select it and click the *I* button on the Formatting toolbar.

If you have to type certain scientific names frequently, use AutoCorrect (see Section 4.6) either to replace the name with a simple keystroke combination or to have it italicized automatically.

4.4 Customizing Your Toolbars

After you have used Word for a while, you may notice that you never use some of the buttons on the toolbars, but you do frequently use commands that are buried deep in the submenus.

Figure A1.27 Customize dialog box, Commands tab

Adding commands to toolbars. You can add frequently used commands, such as super- and subscripts, to the Formatting toolbar. To do this:

1. Click **Tools | Customize** on the menu bar. The **Customize** dialog box has three tabs: Toolbars, Commands, and Options.

2. Select the **Commands** tab (Figure A1.27). The left side shows different categories, such as File, Edit, View and the other categories on the menu bar. The right side shows the commands available for the category selected on the left. These commands are generally the options available in the pulldown menus for the respective menu bar categories.

3. To add the superscript or subscript command to the Formatting toolbar, click the **Format** category (left side of dialog box) and scroll down to the **Superscript** or **Subscript** commands (right side of dialog box).

4. Click the **Superscript** or **Subscript** command and, holding down the left mouse button, drag the command to the desired location on the Formatting toolbar. Instead of having to go through **Format | Font | Select Superscript** or **Subscript** box, you only have to click the appropriate toolbar button.

If you use Word a lot, it may be worth your while to browse the categories and commands in the **Customize** dialog box. Make frequently used commands more accessible by adding them to the toolbar or menu bar, and remove buttons that you never use.

Removing buttons from toolbars. To remove a button from a toolbar, hold down the **Alt** key, click the unneeded button, and drag it away from the menus and toolbars. An *x* appears on the button to show that the command will be removed. If you change your mind later, you can add it back, following the aforementioned instructions.

4.5 Shortcut Keys

It is very time-consuming to go through the **Insert | Symbol | Choose Font | Click symbol | Click Insert** routine every time you want to insert a degree sign in your paper. It is worth your while to make shortcut keys for frequently used symbols. To do this, go to **Insert | Symbol | Font** and click the desired symbol. Notice the **Shortcut Key** button at the bottom of the **Symbol** dialog box (see Figures A1.24 and A1.25). When you click this button, you have the opportunity to define a combination of keystrokes for that particular symbol (Figure A1.28). The combination should be easy

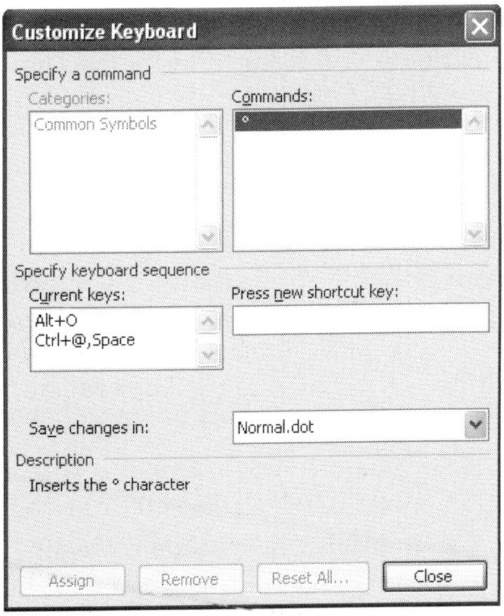

Figure A1.28 Customize Keyboard dialog box

to remember and is usually Ctrl, Alt, Ctrl+Shift, or Ctrl+Alt plus some other character.

Let's say you want to make a shortcut key for the degree sign. The degree sign looks like a lowercase *o*, so you might use Alt+o. You type Alt+o into the **Press new shortcut key** box, and Word notifies you either that this combination is unassigned (as in Figure A1.28) or that it has already been assigned to another command. (*Note:* Even though you typed Alt and a lowercase *o*, the shortcut appears as Alt and upper-case *O* in the box. If you had typed Alt and uppercase *O*, this would have appeared as Alt+Shift+O.)

If the shortcut is already assigned to a different command, determine if you use that command very often. If you don't, you can remove the shortcut from the seldom-used command and assign it to the symbol you want. To do this:

1. Select **Assign** from the buttons at the bottom of the dialog box (see Figure A1.28). Alt+o will then be listed under **Current keys**. If Word assigned a different shortcut key to the degree sign, this will also be displayed under **Current keys** (in this example, Ctrl+@, Space).

2. Click the previously assigned shortcut key, and then select **Remove** from the buttons at the bottom of the dialog box.

3. Close the **Customize Keyboard** dialog box.

4. Close the **Symbol** dialog box.

Next time you have to write the symbol for "degrees Celsius," simply type Alt+o followed by uppercase *C* to get: °C.

4.6 AutoCorrect

AutoCorrect is useful for more than just correcting spelling mistakes. Four possible applications in scientific writing are:

- To replace long chemical names with a simple keystroke combination
- To insert symbols or special characters using just one keyboard key
- To italicize scientific names of organisms automatically
- To insert expressions with sub- or superscripts automatically

Replacing long chemical names with a simple keystroke combination. It is tedious to type units and long chemical names that occur frequent-ly in your text. To save yourself some keystrokes, you can program

AutoCorrect to replace an expression that takes a long time to type with a simple keystroke combination. You must choose the simple keystroke combination judiciously, however, because every time you type these keystrokes, Word will replace them with what you programmed in AutoCorrect. One "trick" is to precede the keystroke combination with either a comma or semicolon (, or ;). Since these punctuation marks are typically followed by a space and not a letter, adding one before your "code letter(s)" creates a unique combination.

For example, let's say "beta-galactosidase" is a word you have to type frequently. You choose ";bg" to designate "beta-galactosidase." To program **AutoCorrect** for this entry:

1. Select **Tools | AutoCorrect** from the menu bar. This opens the **AutoCorrect** dialog box.

2. Type ";bg" in the **Replace** text box.

3. Type "beta-galactosidase" in the **With** text box.

4. Click **Add** button.

5. Click **OK**.

Inserting symbols or special characters using just one keyboard key. If you want to use AutoCorrect to insert symbols or special characters, the procedure is as follows:

1. Select **Insert | Symbol** from the menu bar. This opens the **Symbol** dialog box (see Figure A1.24).

2. Change the font if the symbol you want to use is not in the (normal text) font.

3. Click the symbol.

4. Click the **AutoCorrect** button at the bottom of the dialog box. The symbol will already be entered in the **With** text box.

5. In the **Replace** text box, type the characters you want to replace with the symbol.

6. Click **Add** button.

7. Click **OK**.

Italicizing scientific names of organisms automatically. If you are going to be typing the scientific name of an organism repeatedly, you can insure that it always appears in italics without manually imposing italics each time. To do this, follow these steps:

1. First type the scientific name of the organism in your text (e.g., Aphanizomenon flos-aquae).

2. Italicize the name by selecting it and then clicking the **Italicize** button (*I*) on the Formatting toolbar or typing the keyboard command Ctrl+I; either way, the text should become *Aphanizomenon flos-aquae*.

3. With the name still selected, click **Tools | AutoCorrect** on the menu bar.

4. On the **AutoCorrect** tab, you will notice Aphanizomenon already entered in the **With** text box. Click the **Formatted text** option button to italicize it.

5. Enter a shorter version of the name in the **Replace** text box, for example "aphan."

6. Click the **Add** button.

7. Click **OK** to close the window.

Inserting expressions with sub- and superscripts. The principle of using AutoCorrect to insert expressions with sub- and superscripts is the same as that for italicizing scientific names.

1. Type the expression without sub- or superscripts (e.g., Vmax).

2. Select the characters to be sub- or superscripted, and then format them as described in Section 4.2 or 4.4 (the expression then becomes V_{max}).

3. Select the entire expression, and click **Tools | AutoCorrect**.

4. On the AutoCorrect tab, you will notice "V_{max}" already entered in the **With** text box. Click the **Formatted Text** option button to add the subscripting.

5. Type "Vmax" in the **Replace** text box.

6. Click the **Add** button.

7. Click **OK** to close the window.

4.7 Tables Inserted in the Document

Depending on the guidelines of the scientific journal, tables may either be attached on separate pages at the end of the paper or may be inserted as soon as feasible following the text where the table is first mentioned. Ask your instructor if he or she has a preference. Most people find it easier to refer to a table that is in close proximity to the text.

Word makes it easy to create tables to your specifications (see Section 2.12). You can also make tables in Excel (see "Importing Tables From Excel Into a Word Document" in Appendix 2), and then import and format them in Word.

4.8 Figures Imported from Another Program Such as Excel

As with tables, figures may either be attached on separate pages at the end of the paper or may be inserted as soon as feasible following the text where the figure is first mentioned. Ask your instructor if he or she has a preference. Most people find it easier to refer to a figure that is in close proximity to the text.

The general idea behind importing figures from another program into Word is:

1. Activate the desired graphic.
2. Click the **Copy** button on the Standard toolbar (or **Edit | Copy** on the menu bar) in the graphing program.
3. Position the cursor at the location in the Word document where you would like to insert the graphic.
4. Click the **Paste** button on the Standard toolbar (or **Edit | Paste** on the menu bar) in Word.

To import a graph (figure) from Excel, follow these steps:

1. First you must make the graph in Excel (see Appendix 2). **Make all formatting changes in Excel,** as it is not possible to make any changes (except to the size) once the graph has been imported into your Word document.
2. When you are satisfied with the graph in Excel, single-click the area outside the axes, but inside the frame (**Chart Area**, not **Plot Area**).
3. Still in Excel, click the **Copy** button on the Standard toolbar.
4. Position the cursor at the desired location in your Word document.
5. Now in Word, click the **Paste** button on the Standard toolbar.
6. The figure will appear in the Word document.
7. Type a figure caption **below** the figure. The caption consists of the word "Figure" followed by a number and descriptive title. Center the caption or align it flush on the left margin. Use Arabic numbers and number the figures consecutively in the order they are described in the text.
8. Figure titles should be informative and consist of a precise noun phrase (not a complete sentence). Titles that merely state the y-axis legend versus the x-axis legend are not acceptable.

4.9 Equations

It is unlikely that you will need to type complex mathematical equations in a scientific paper for an introductory biology course. If you are interested in doing so, however, you can open Microsoft Equation Editor by selecting **Insert | Object**. Select **Microsoft Equation 3.0** as the object type. Several references in the Bibliography provide instructions.

4.10 Leaving Space for Sketches

In some laboratory exercises, you may be asked to sketch cells, tissue sections, or specimens that you viewed under the microscope. By convention, these sketches are called "figures" and require a figure caption (number and title) beneath the figure.

It is difficult to judge on the computer screen how much space to leave for your sketches. You don't want to leave too small a space, because your sketches should be large, clear, and with some detail. On the other hand, you don't want to leave too much space, because then it looks like something is missing.

To measure space precisely, invest in a word processor's ruler, which you can buy at any good office supply store. This ruler has a scale that is divided into tenths of an inch, the same units used to tell you the distance of the cursor from the top of the page (see Status bar at the bottom of the screen).

One technique to determine how much space to leave for your sketch is shown in Figure A1.29.

Step-by-step instructions for this method are as follows:

1. First make a rough draft of your sketch so that you have an idea of the space requirement (Figure A1.29A).
2. Type the first draft of your scientific paper (Figure A1.29B).
3. Determine where you want to insert the sketch in your text. It is preferable to insert the figure as soon as possible after the paragraph where it is first described.
4. Position the cursor on the last line of the paragraph in which the figure is described. Note the distance of this line from the top of the page (in inches), as indicated on the status bar at the bottom of the screen. In Figure A1.29B, this is 1.5".
5. Lay the ruler on your sketch so that the mark for the line distance determined in step 4 is placed just above the top of your sketch.
6. Note the line distance on the ruler just beneath your sketch. In Figure A1.29A, this is 6.8". This number represents the line distance at which you will type the caption for your sketch. If this

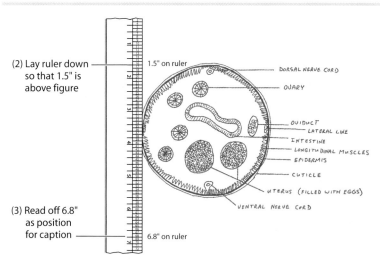

(2) Lay ruler down so that 1.5" is above figure

1.5" on ruler

DORSAL NERVE CORD
OVARY
OVIDUCT
LATERAL LINE
INTESTINE
LONGITUDINAL MUSCLES
EPIDERMIS
CUTICLE
UTERUS (FILLED WITH EGGS)
VENTRAL NERVE CORD

(3) Read off 6.8" as position for caption

6.8" on ruler

(A) Rough sketch of cross section of female *Ascaris*, 40x

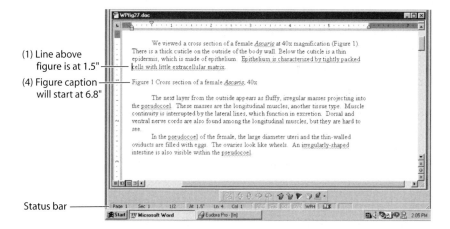

(1) Line above figure is at 1.5"

(4) Figure caption will start at 6.8"

Status bar

(B) Computer screen displaying text into which you plan to insert a sketch

Figure A1.29 A technique for measuring space for hand-drawn sketches. After creating the sketch (A) on paper and typing your document on the computer (B), take the following steps. (1) In the document, position the cursor just above where you want to place the sketch. Look at the status bar at the bottom of the screen to determine this distance from the top of the page (in inches). In this example, the distance is 1.5". (2) Lay a ruler on your sketch so that 1.5" is about 0.5" *above* the top of your sketch. (3) Read the distance (in inches) at the point about 0.5" *below* the bottom of your sketch. In this example, that distance is 6.8". (4) Insert returns in your document until the status bar shows that the cursor is at the distance you determined in (3). Type the figure caption on this line.

number is greater than roughly 9.6″, you will have to place your sketch on the next page, because there is not enough room for it on the current page.

7. If there is not enough room for the sketch at the bottom of the page, use your judgment to decide whether or not to type more text after the paragraph in which your sketch was first mentioned. If you decide not to type more text, insert a page break, and continue with the text after the caption for your sketch.

4.11 Hanging Indent to List References in the Name-Year System

The Council of Science Editors (CBE Manual, 1994) recommends two formats for listing references, the Citation-Sequence System and the Name-Year System (see Chapter 4, pp. 59–63).

If you use the Citation-Sequence system, the references are numbered according to order of citation. See Section 2.13 of this appendix for instructions on how to make numbered lists.

If you use the Name-Year system, the references are listed in alphabetical order according to the first author's last name. The first line of the reference begins on the left margin, and the subsequent lines are indented. This style is called a **hanging indent**. References typed with hanging indent format look like this:

Murata T, Kadota A, Wada M. 1997. Effects of blue light on cell elongation and microtubule orientation in dark-grown gametophytes of *Ceratopteris richardii*. Plant and Cell Physiology 38(2): 201–209.

There are several ways to apply hanging indent format to the list of references. Perhaps the easiest is to type each reference so that it is aligned on the left margin, with an **Enter** (Return) at the end of each reference. When you are finished, select all the references (see Section 2.7), and then click **Format | Paragraph** on the menu bar. On the **Indents and Spacing** tab (see Figure A1.10), scroll down to **Hanging** in the **Indentation: Special** box. You can set the indentation to 0.25″ or 0.5″ in the **By:** box.

MAKING XY GRAPHS IN MICROSOFT EXCEL

Graphs are characteristic of scientific and technical writing. They help the reader to visualize the data, and to see relationships among variables. Graphs are a way to summarize information that would otherwise involve lengthy descriptions that would be difficult to interpret.

Years ago, authors used drawing instruments, specially lined graph paper, and ink to make figures for published papers. Mistakes were difficult, if not impossible to correct, so any mistake usually resulted in starting over. If a different sized figure was required, the entire figure had to be redrawn. The advent of plotting software in the mid-1980s, however, has made hand-drawn graphs practically obsolete. You simply enter the data into a graphics program in a prescribed manner and select the desired form for the graph; the computer draws the graph for you.

The most common types of graphs are **line graphs**, **bar graphs**, and **pie charts**. The type of graph you choose should be based on which one best depicts and emphasizes the trends shown by your data. Some guidelines for helping you choose an appropriate type of graph are given in Chapter 4.

This appendix gives you step-by-step instructions for making XY graphs in Microsoft Excel. Excel is a good plotting program for novices for the following reasons:

- Data input and subsequent plotting of these data is relatively straightforward in Excel.

- Excel is readily available and is included in the Microsoft Office suite of computer software.

- If your school has Excel on its computers, it's likely that you can get assistance from a staff member in your school's computer services department.

The time you invest now in learning to plot data on the computer will be invaluable in your upper-level courses and later in your career. You may eventually switch to a higher-powered plotting program such as SigmaPlot, but the experience gained by working with Excel should make this transition easier.

The format described in this appendix follows the Council of Science Editors' recommendations (CBE Manual, 1994). Instructions to "single-click" or "double-click" refer to clicking the *left* mouse button, unless otherwise noted.

The screen drops in this appendix are for Word 2003, but the instructions apply to earlier versions as well, unless otherwise noted. Excel commands are distinguished in this appendix with a vertical bar, separating commands from subcommands. For example, **Tools | Customize | Options** means "Click the Tools button (on the menu bar), then select Customize, and then select the Options command."

The Microsoft Excel Screen

Click the **Excel** button on the taskbar at the bottom of your screen to open an Excel worksheet (spreadsheet). Alternatively, find Microsoft Excel under **Start | All Programs | Microsoft Office | Microsoft Office Excel**. A worksheet consists of thousands of cells arranged in rows and columns, with menus and toolbars located above the cells at the top of the screen display (Figure A2.1). The columns have letter headings (A, B, C, etc.) and the rows have numerical headings (1, 2, 3, etc.).

Title bar

The title of the worksheet is given at the very top of the page on a blue background. If you have not yet named the worksheet, the title bar will read **Microsoft Excel–Book 1**.

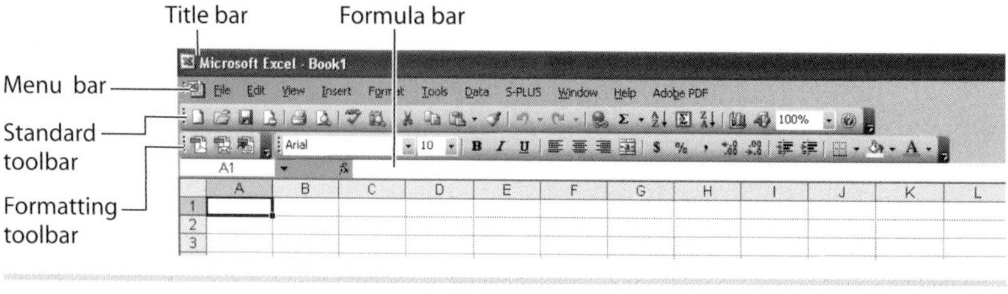

Figure A2.1 Screen display of Excel 2003 worksheet

Excel 2003 Update

The Microsoft Office XP programs Excel 2002 and Excel 2003 look different from their predecessors. The new default window is split into the **Document** pane and the **Task** pane (Figure A2.2).

The Task pane appears when you choose certain commands or when you click **View | Task pane** on the menu bar. To switch to another Task pane, click the title bar at the top of the Task pane. To close the Task pane, click the **Close** box (⊠) in the top right corner of the Task pane.

Menu bar

Just below the title bar is the menu bar. The menu bar consists of buttons that, when clicked, list other commands in pulldown menus. For example, when you click **File**, the pulldown menu shown in Figure A2.2 is displayed.

Toolbars

If your Excel program has been set up to display toolbars (Standard and Formatting), they are displayed just below the menu bar. Toolbars contain pictorial buttons that are shortcuts to menu commands.

If the Standard and Formatting toolbars are not displayed, go to the menu bar and select **View | Toolbars** to see if there is a checkmark next to **Standard** and **Formatting**. If there is no checkmark, go to the last item on the Toolbars pulldown menu and click **Customize**. Click the boxes next to **Standard** and **Formatting**, then click **Close**.

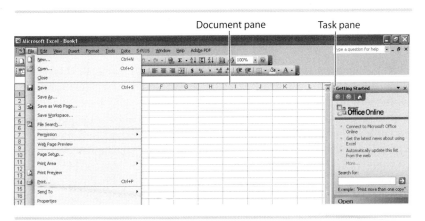

Figure A2.2 Pulldown menu for File command

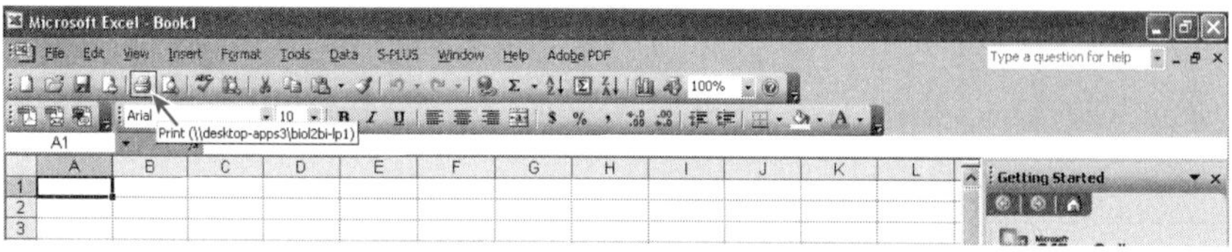

Figure A2.3 Selecting the Print button on the Standard toolbar

When you place the mouse pointer over a particular button on the toolbar or menu bar (do not click any mouse buttons), a description of that button will appear. For example, if you place the mouse pointer on the 5th button from the left on the Standard toolbar (Figure A2.3), the **Print** command, with the default printer in parentheses, will appear in a pop-up label. The **Print** command is also found on the menu bar under **File | Print**.

Note: Shortcuts offer only one option from the corresponding menu command. For example, if you click the **Print** button on the Standard toolbar, the entire document will be printed on the default printer. If you want to print single pages or use a different printer, you must go to the menu bar and select the desired options from the **File | Print** pull-down menu.

Formula bar

The formula bar is located just below the toolbars. It is used to enter formulas for performing calculations on the data entered in the spreadsheet. It is also used to edit the contents of a cell.

Entering Data in Spreadsheet

Before you enter data in an Excel worksheet, you must first have a clear idea of what your XY graph should look like. Which parameter should be plotted on the x-axis and which one on the y-axis? By convention, the x-axis of the graph shows the independent variable. The independent variable is the one that the scientist manipulated during the experiment. The y-axis of the graph shows the dependent variable, the variable that changes in response to changes in the independent variable.

Let's say you conducted an experiment in which you determined how the activity of an enzyme (catalase) changed when you varied the temperature. Because you manipulated the temperature in the experiment,

	A	B	C	D	E
1					
2					
3					
4					
5					

TABLE A2.1
Portion of Excel spreadsheet

temperature is the independent variable and should be plotted on the x-axis. Catalase activity is the dependent variable because it changes in response to changes in the temperature, and should be plotted on the y-axis.

On the Excel spreadsheet, **Column A** is used to enter the data for the x-**axis**, whereas subsequent columns are used for data for the y-**axis** (Table A2.1). In the previous example, temperature data would be plotted in Column A and the corresponding catalase activity data would be plotted in Column B.

Enter the data as in Table A2.2. For a simple graph with only one data set, it is not necessary to provide column headings.

TABLE A2.2 Sample data for effect of temperature on catalase activity. Temperatures (°C) are entered in Column A, catalase activity (units of product formed $\cdot$ sec^{-1}) in Column B.

	A	B
1	4	0.039
2	15	0.073
3	23	0.077
4	30	0.096
5	37	0.082
6	50	0.04
7	60	0.007
8	70	0
9	100	0

TABLE A2.3 Excel-specific terminology

EXCEL TERM	DESCRIPTION
Chart	Graph
Category axis	x-axis
Value axis	y-axis
Data series	Set of related data points
Plot area	Area of the graph inside the axes
Chart area	Area outside the axes but inside the frame
Legend	Legend or key

Terminology

The terminology that Excel uses for graphs is different from what biologists use. For example, you have already learned that Excel calls a spreadsheet a worksheet. The Excel instructions provided here use the terms biologists use, rather than those chosen by Excel. It is helpful to know the Excel terminology when you use the Help menus, however, and Table A2.3 translates some common Excel terms into biologists' terms.

Using Chart Wizard to Plot the Data

1. In the spreadsheet, **select the data to be plotted using the mouse**. To do this, single-click the left mouse button on the first cell, and keep holding the button down as you drag the mouse pointer over the rest of the cells that contain the data to be plotted. When you release the mouse button, the cells you selected are highlighted in blue. Alternatively, click the top left box of the block of data, hold down the **Shift** key, and click the bottom right box of the block. All the selected cells are highlighted in blue.

   Note: If the data to be plotted are not in adjacent columns, highlight the first column, hold down the **Ctrl** key, and then highlight the data in any other column(s).

2. **Click Chart Wizard on the Standard toolbar.** The **Chart Wizard** button is a colored bar graph (Figure A2.4). Using four sequential dialog windows, Chart Wizard walks you through the steps needed to make a first draft of your graph.

3. Chart Wizard Step 1 of 4—Chart Type allows you to **select the type of plot** (Figure A2.5).

Figure A2.4 Selecting Chart Wizard on the Standard toolbar

- To make a line graph, select **XY (Scatter)**. Do *not* select **Line**, because this option spaces the *x*-axis values at equal intervals, instead of according to the intervals of the data.
- For Chart sub-type, you must decide whether the points should be connected or not (see Chapter 4 for an explanation of these formats). If there are only a few data points, and you want to show the relationship between the independent variable and the dependent variable, then select **Data points connected by smoothed or straight lines**. Usually your instructor wants to see the data points, so do *not* select the options in which Excel plots the curve without

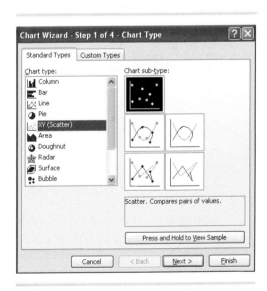

Figure A2.5 Selecting a type of graph in Chart Wizard, Step 1 of 4. Current selection is XY Scatter with unconnected data points.

displaying the data points. If you are making a standard curve or a scatter plot, then select **Scatter with no points connected**.

- Click **Next**.

4. Chart Wizard Step 2 of 4—Chart Source Data tells you which cells you selected on the spreadsheet. **The data series should be in Columns**. Click **Next**.

5. Chart Wizard Step 3 of 4—Chart Options shows you five tabs: **Titles, Axes, Gridlines, Legend**, and **Data Labels** (Figure A2.6).

   📁 Titles

   Chart Title: Leave blank if this graph is for a lab report or poster. By doing so, Excel will make a larger graph for you to import into your text document later. In the text document, type the figure caption *under* the figure. **Add a title** if the graph is for an oral presentation.

   Value (X) axis: Enter the x-axis label with the units in parentheses. For the data in Table A2.2, the x-axis label would be: Temperature (°C). For the time being, don't type the degree sign. You can copy and paste it later (see the section on "Format Axis Label").

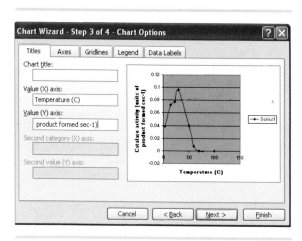

Figure A2.6 Dialog box in Chart Wizard, Step 3 of 4 for entering title and axis labels and hiding or displaying gridlines and legends

Value (Y) axis: Enter the *y*-axis label with the units in parentheses. For the data in Table A2.2, the *y*-axis label would be: Catalase activity (units of product formed • sec^{-1}). For the time being, leave out the raised dot and simply type "–1" for the exponent.

📁 **Axes**

No changes.

📁 **Gridlines**

There should be no gridlines on the figure. Clear the box next to Value (Y) axis: Major gridlines.

📁 **Legend**

No legend (key) is required when there is only one line on the graph. Clear the box next to Show legend.

📁 **Data labels**

None.

When you have finished with these five tabs, click **Next**.

6. Chart Wizard Step 4 of 4—Chart Location asks you if you want the figure to be printed as a new sheet or as an object in the current sheet.

   ▪ By convention, figures are placed as close as possible to the location in the scientific paper where they are first described. Because you have to import the graph into a text document anyway, it does not really matter whether you choose **As object in Sheet 1** or **As new sheet**. Selecting **As object in Sheet 1** has the advantage that graph size is uniform when the object is imported into the text document. You can, however, adjust the size of the graph in the text document with either option.

   ▪ If the figure will be mounted on a poster or attached as a separate sheet at the end of the lab report, then select **As new sheet**.

   ▪ Click **Finish**. Your graph should look like Figure A2.7.

7. Inspect the graph carefully. Look for proper units and spacing on the *x*- and *y*-axes, for appropriate labeling of the axes, and for expected trends of the line itself. If you notice that you entered a datum incorrectly in the spreadsheet, simply change it, and the correction will also be made in the graph.

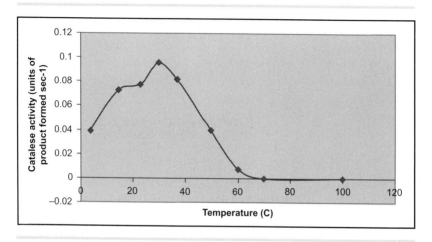

Figure A2.7 First draft of Table A2.2 data plotted by Chart Wizard. See Figure A2.13 for the final form.

Modifying Graphs

Graphs produced by Chart Wizard can be modified to meet an instructor's or a journal's specifications. The modifications described in this section follow the guidelines given in the CBE Manual (1994). To modify a graph, first activate it by single-clicking the plot area or chart area (Figure A2.8). Then click the **Chart** button on the menu bar. You can return to any of the four steps of Chart Wizard (Chart Type, Source Data, Chart Options, and Location), or you can select **Add Data** (type in range of new data to be plotted from spreadsheet) or **Add Trendline**.

In Excel 2000 and later versions, when you single-click the plot area or chart area, a floating toolbar with **Chart** commands appears on the screen (Figure A2.9). If it doesn't, click **View | Toolbars** and check the box next to **Chart**. Select any of the **Chart** options in the pulldown menu, then click the **Format** button to change the format of that particular option. There are more options under the **Chart** command on the menu bar than on the floating toolbar.

Note: The **Chart** command will not be displayed on the menu bar unless you first activate the graph (click inside the chart or plot area to activate the graph).

Chart Type

If you want to see what the data look like plotted in another form, select **Chart | Chart Type**. Select a different type of graph, and click the **Press**

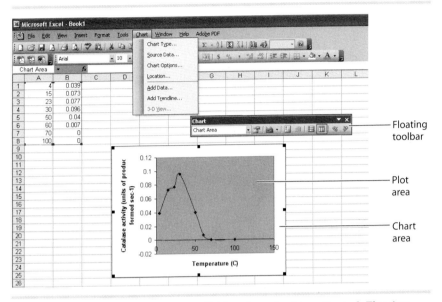

Figure A2.8 Formatting options available when chart is activated: Floating toolbar or Chart button on the menu bar

and Hold to View Sample button. Check with your instructor regarding what chart type is appropriate.

Source Data

If you want to change the range of the cells used to plot the graph, you can do so in this dialog window. This might be appropriate if you added or deleted data values in the spreadsheet.

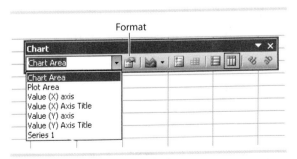

Figure A2.9 Floating toolbar options for formatting different parts of the graph

Chart Options

This allows you to change the x- and y-axis labels and delete gridlines and the legend. The **Axes** tab seems to suggest that the format of the axes can be changed here, but this is misleading. For more information, see the section on "Format Axes."

Location

If you expect to insert the figure into a Word document (this is preferred when writing a lab report), place the figure as an object in the worksheet. If you plan to mount the graph on a poster or attach it as a separate sheet at the end of the document, select **As new sheet**.

Add Data

To add data to an existing figure, first enter the new data in cells on the spreadsheet. Then click the chart area of the existing figure, select **Chart | Add Data**, and change the range of cells.

Add Trendline

A trendline is a line that describes the general tendency of the data points. In Excel you can add any of six trendlines or regression lines (best-fit lines) to a data set. Excel automatically calculates the R-squared value and the equation of the line. The equation allows you to predict one variable when the other is known. The R-squared value gives you an indication of how well the line fits the data points (the closer to 1, the better the fit). See the section on "Trendlines" for situations when trendlines may be added to a data set.

Format Data Series

Excel's hierarchy for symbols used in multiple line graphs is not the same as that recommended by the Council of Science Editors (CBE Manual, 1994). The two hierarchies are compared in Table A2.4.

The symbols that science editors use are based on ease of recognition and good contrast in black and white journal publications. The criteria used by Excel to determine symbol hierarchy are not obvious, especially since light colored symbols are very hard to see on a white background.

To change the symbol style and/or color in Excel, double-click the line or the symbol to open the **Format Data Series** window (Figure A2.10). On the **Patterns** tab:

TABLE A2.4 Comparison of Excel and CBE Manual symbol hierarchy for line graphs

EXCEL	CBE MANUAL
Navy blue diamond	Black open circle
Pink square	Black filled circle
Yellow triangle	Black open triangle
Turquoise x	Black filled triangle
Purple x with additional vertical line	Black open square
Brown diamond	Black filled square

- Change **Line Color** from Automatic to Black.
- Change the **Marker** (symbol) to a circle, triangle, or square by selecting the desired symbol next to **Style**. Next to **Foreground**, choose **Black**. Next to **Background**, choose **No color** to make an open circle, triangle, or square; or **Black** to make a filled circle, triangle, or square.

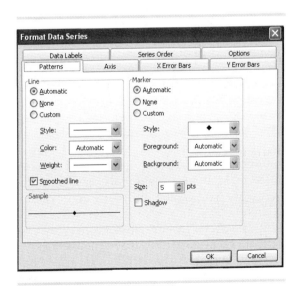

Figure A2.10 Patterns tab inside Format Data Series dialog box, opened by double-clicking a data series (line or symbol) on a line graph

Format Plot Area

Traditionally, scientific journals are published in black and white. With this in mind, make the background color (plot area) **white** for best contrast. To select the background color, double-click the plot area (area inside the axes). In the **Format Plot Area** dialog box, set the **Area** to None. Set the **Border** to None. Click **OK**.

Format Chart Area

Graphs in scientific journals do not have a border. To remove the border that Excel automatically creates around the figure, double-click the chart area (area outside the axes, but inside the frame). In the **Format Chart Area** dialog box, set the **Border** to None. The default for the **Area** (white) is fine.

Format Legend

The background color of the legend (key) is usually white and there is no border. Double-click the legend to open the **Format Legend** dialog box. On the **Patterns** tab, set the **Border** to None and leave the default selection for **Area**: Automatic (white). Click **OK**.

When possible, place the legend within the plot area. This is done by single-clicking the legend (selection handles will appear) and dragging it to the desired location.

Format Axes

Use the criteria in Table A2.5 to check that the axes have been formatted correctly. If any changes need to be made, double-click any number on either the x- or y-axis (depending on which axis needs to be changed). This action opens the **Format Axis** dialog box (Figure A2.11). Five tabs are displayed: **Patterns**, **Scale**, **Font**, **Number**, and **Alignment**. Click the appropriate tab (see Table A2.5) and make the necessary changes.

Format Axis Label

Use the criteria in Table A2.6 to check that the x- and y-axis labels have been formatted correctly. If any changes need to be made, double-click the appropriate label. This action opens the **Format Axis Title** dialog box (Figure A2.12). Three tabs are displayed: **Patterns**, **Font**, and **Alignment**. Click the appropriate tab (see Table A2.6) and make the necessary changes.

To **superscript** or **subscript** certain characters in the x- and y-axis labels, or in the legend, single-click the text box to activate the label. Use

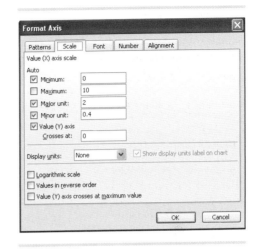

Figure A2.11 Scale tab inside Format Axis dialog box, opened by double-clicking any number on either the *y*-axis or the *x*-axis

the mouse to select the character(s) to be super- or subscripted, and then select **Format | Selected Axis Title** on the menu bar. On the **Font** tab, select either super- or subscript in the **Effects** category. Click **OK**.

Inserting Greek Characters and Other Symbols in Axis Labels

Greek characters and **other symbols** can be inserted into axis labels by using two straightforward methods: copying and pasting the symbol or character from a Word document, or by selecting the symbol from the Character Map.

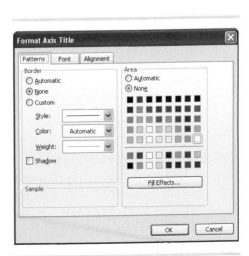

Figure A2.12 Patterns tab inside Format Axis Title dialog box, opened by double-clicking either the *y*-axis or the *x*-axis label

TABLE A2.5 Checklist for axis format

CORRECT FORMAT	HOW TO ADJUST
Dependent variable on *y*-axis	See "Entering Data in Spreadsheet"
Independent variable on *x*-axis	See "Entering Data in Spreadsheet"
Range of values on axes is slightly larger than range of data values being plotted	Scale tab | Min, Max If negative values are not appropriate for the quantity plotted, Min = 0.
Axis numbers are multiples of 2, 5, or 10 whenever possible	Scale tab | Major unit Minor units are not displayed, so leave the default selection.
If scale includes axis numbers less than 1, a zero is required before each decimal point	See "Entering Data in Spreadsheet"
Tick marks are left of the *y*-axis and below the *x*-axis	Patterns tab | Major tick mark type: Outside
Tick marks should always be accompanied by a number (no subdivisions between numbered marks)	Patterns tab | Minor tick mark type: None
Numbers should be centered on their respective tick marks, outside the field of the graph	Patterns tab | Tick mark labels: Next to axis
Numbers should be the same size and should read horizontally	Alignment tab | Orientation | Automatic

Copy and Paste from Word document

1. Type the desired symbol in Microsoft Word (see Appendix 1, Section 4.1 or 4.5).
2. Select the symbol and click **Edit** | **Copy** on the menu bar.
3. In your Excel file, single-click the axis label that contains the special symbol. Put the cursor where you want to insert the symbol. Click **Edit** | **Paste**.

Select Symbol from Character Map

1. For PCs, click the **Start** button located in the lower left corner of your screen. Select **All Programs** | **Accessories** | **System Tools** | **Character Map**.
2. In the Font section located at the top of the **Character Map** window, scroll down to **Symbol** and highlight it.

TABLE A2.6 Checklist for axis label format

CORRECT FORMAT	HOW TO ADJUST
Capitalize only the first word of the label and any proper nouns	Retype
A word or a phrase that accurately describes the variable	Retype
Centered on the length of the axis	Alignment tab | Text alignment Horizontal: Center Vertical: Center Orientation: 0 degrees
Vertical axis labels read parallel to the y-axis (they should never read vertically downward)	Alignment tab | Text alignment Vertical: Center Orientation: 0 degrees
Units of measurement for the variable are placed in parentheses after the variable	Retype
Font typeface and size	Font tab Font: Arial or another *sans serif* font Font style: Regular or bold Size: At least 10pt
Superscripted or subscripted characters	See below
Special characters (e.g., Greek letters)	See below
No borders or background	Patterns tab Border: None Area: Automatic (white)

3. Locate the symbol you want to include on the axis label, and single-click it. Let's use μ (Greek mu) as an example.

4. Click the **Select** button and then the **Copy** button at the bottom of the **Character Map** window.

5. In your Excel file, single-click the axis label that contains the special symbol. Put the cursor where you want to insert the symbol. Click **Edit | Paste**.

6. Use the mouse to highlight the "m" that you want to change to μ.

7. On the formatting toolbar, click the arrow button next to the **Font** box and select **Symbol**.

Applying CBE Guidelines to Figure A2.7

The criteria listed in the previous section can now be used to evaluate the graph produced by Chart Wizard (see Figure A2.7). Problems with the graph and how to correct them are summarized in Table A2.7. The final form of the graph is shown in Figure A2.13.

Make sure you make all of the necessary modifications and corrections before you copy the graph into your text document (see the next section on "Importing Graphs into Microsoft Word") or before you print it out.

Importing Graphs into Microsoft Word

In order to give your lab report a professional appearance, insert figures as close as possible to the location in the document where they are first mentioned. This can be done easily by copying the graph you made in Excel and pasting it into a Microsoft Word document (this method also works for WordPerfect documents).

Here are some tips that will help you carry out this task:

1. **Make all modifications and corrections to the graph in Excel.** Once it is imported into the Word document, it is no longer possible to make any changes (except to the size of the graph).

Catalase activity peaked at 30°C (Figure 1). From 4 to 30°C, activity steadily increased to a maximum of 0.096 units of product formed • sec⁻¹, and then activity fell with increasing temperature until there was no activity beyond 70°C.

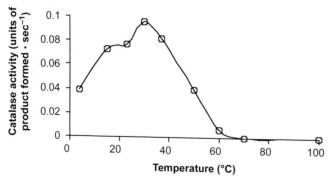

Figure 1 Effect of temperature on catalase activity

Figure A2.13 Excerpt from a Results section in which the description of the results precedes the figure, the figure is referenced parenthetically, and the figure is formatted properly

TABLE A2.7 Problems with Figure A2.7 and how to remedy them

PROBLEM	REMEDY
Plot area (background) should be white, not gray; plot area should not have a border	Double-click plot area to open Format Plot Area dialog box. Select: Border: None Area: None or white
Chart area should not have a border	Double-click chart area to open Format Chart Area dialog box. Select: Border: None Area: Default (white) or none
Symbol is a blue diamond; it should be a black open circle	Double-click data series (line or symbol) to open Format Data Series dialog box. Select: Line Color: Black Marker Style: Circle Foreground: Black Background: None (for an open symbol)
Range of values on axes 1. x-axis: 0 to 120 is 20% larger than needed 2. y-axis: Negative rates don't make sense. Minimum value should be zero. Maximum value should be 0.1.	Double-click any number on the respective axis. 1. On the Scale tab for the x-axis, change Max to 100. 2. On the Scale tab for the y-axis, change Min to 0 and Max to 0.1.
Axis labels 1. Font size: at least 10 pt 2. x-axis: Insert a degree symbol (°) 3. y-axis: Superscript −1 4. y-axis: Insert raised dot (·)	Single-click the respective axis label. 1. Format | Selected Axis Title on the menu bar. On the Font tab, Size: 10 pt or larger 2. Insert a degree symbol in Word (Insert | Symbol | °), then Copy and Paste it into the x-axis label; alternatively, in Excel, locate degree symbol in Character Map, then Copy and Paste (see "Format Axis Label"). 3. Select −1. Then click Format | Selected Axis Title. On the Font tab, Effects: Superscript. Note: The y-axis label will temporarily be turned 90° so that it reads horizontally. When you finish with the formatting, click outside the figure and the label will return to its original orientation. 4. See (2).

2. To copy the graph, single-click the Chart Area (area inside the frame but outside of the axes). Selection handles will be displayed on the frame, indicating that the graph is activated. *Note*: If you mistakenly *double-click* the chart area, the Format Chart Area window comes up. Click **Cancel** to close the window.

   When the graph is activated, click the **Copy** button on the Standard toolbar (or **Edit | Copy** on the menu bar). The frame will then have moving dashes.

3. Move the cursor to the location in the Word document where you want to insert the graph. Click the **Paste** button on the Standard toolbar (or **Edit | Paste** on the menu bar). The graph will be pasted into the document at the position of the cursor. If you pasted the graph in the wrong place, click the **Undo Typing** button on the Standard toolbar or **Edit | Undo Typing** on the menu bar. If you notice an error in the graph, make the correction in Excel, and then copy and paste the graph again.

4. See Figures 4.1 and 4.4 in Chapter 4 for guidelines on how to refer to the figure in the text of the Results section of the report or paper and what constitutes a suitable figure caption.

Figure A2.13 shows the final form of the graph incorporated in the Results section of the report or paper.

Trendlines

Trendlines (also called regression lines or "best-fit" lines) are used to look for patterns in a data set and to make predictions about one variable when the other variable is known.

If you are looking for a pattern in a data set, first plot the data as a scatter plot (**Chart Wizard | Step 1 of 4 | XY (Scatter) | No points connected**). Figure A2.14A, for example, shows class data for a lab exercise in which we would like to know if there is a correlation between caffeine concentration and the change in pulsation rate of blackworms. While there is a significant variation in response for any given caffeine concentration, it looks like a bell-shaped curve might describe the pattern. In Figure A2.14B, a second-order polynomial trendline was fitted to the scatter plot by activating the chart and then clicking **Chart | Add Trendline | Type tab: Polynomial, 2nd order** and **Options tab: Display equation on chart** and **Display R-squared value on chart**. The degree to which the trendline actually fits the data is given by the R-squared value,

(A)

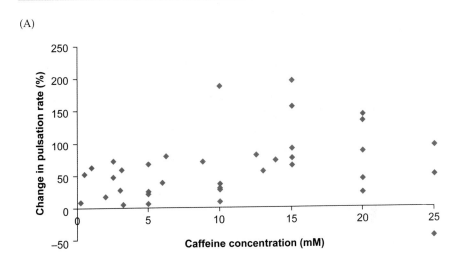

(B)

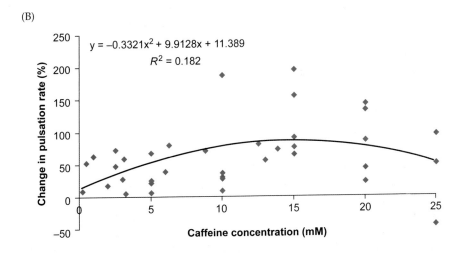

Figure A2.14 Effect of caffeine on the pulsation rate of blackworms. The percent change was determined by subtracting the rate before treatment from the rate after treatment, dividing by the rate before treatment, and multiplying by 100. (A) Data initially graphed as a scatter plot to look for possible patterns. (B) The same data with a second-order polynomial trendline fitted by Excel. The low R^2 value indicates that this trendline does not describe the correlation between change in pulsation rate and caffeine concentration very well.

whereby the closer the R-squared value is to 1, the better the correlation or fit. The low R-squared value in this example indicates that the second-order polynomial trendline does not describe the correlation between dose and response very well. We might conclude from these data that caffeine is a stimulant (because pulsation rate mostly increased at all caffeine concentrations tested), but that there is no clear relationship between caffeine concentration and response.

If your objective is to make predictions about one variable when the other is known, you may try to fit a linear trendline to the data set, as for a standard curve. Some examples in biology where standard curves are used include:

- Predicting the protein concentration of an unknown sample using the correlation between absorbance and known protein concentration (e.g., Biuret method, Bradford method).

- Determining the size of a DNA fragment by comparing the distance it migrated on a gel to the distance migrated by DNA fragments whose sizes are known.

- Quantifying nitrogen or phosphorous in aquatic ecosystems.

Do not insert a linear trendline when the data points do not fall on a straight line! If the data are not being used to make a prediction, just connect the points with straight or smoothed lines (**Chart Wizard | Step 1 of 4**, select **XY (Scatter) | Smoothed lines** or **Straight lines**).

Example: Standard Curve for a Protein Assay

Let's say you did an experiment in which you used the Biuret method to determine protein concentration in an egg white. In the first step of the procedure, you measured the absorbance of four known concentrations of bovine serum albumin (BSA) to make the standard curve. Because of Beer's Law, we expect the standard curve—a graph of absorbance vs. protein concentration—to be a straight line. Follow these steps to plot the standard curve:

1. Enter the BSA concentration in Column A and the corresponding absorbance values in Column B (Table A2.8).
2. Select the cells containing the data and click the **Chart Wizard** button. In Step 1 of 4, select **XY (Scatter) | No points connected**.
3. After completing the rest of the steps in Chart Wizard (see pp. 180–183), and modifying the background, borders, x-axis scale,

TABLE A2.8 Sample data for Biuret standard curve. Protein concentrations (mg/mL) are entered in Column A, absorbance values (at 550 nm) in Column B.

	A	B
1	2	0.10
2	3	0.12
3	5	0.24
4	10	0.40

and symbols according to CBE Manual guidelines, the resulting graph should look like Figure A2.15.

4. Single-click any point to activate the unconnected data set. All the data points should now be highlighted (yellow).

5. Select **Chart | Add Trendline** on the menu bar.

6. There are two tabs under the **Add Trendline** option: **Type** and **Options**. On the **Type** tab, select the default: **Linear**.

7. On the **Options** tab, check the boxes next to **Display equation on chart** and **Display R-squared value**. The equation allows you to predict the concentration in a sample of egg white (the "unknown") when its absorbance is known (measured). The R-squared value tells you how well the equation fits the data.

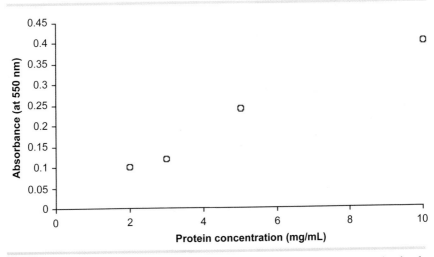

Figure A2.15 Biuret standard curve as a scatter plot. See Figure A2.16 for final form.

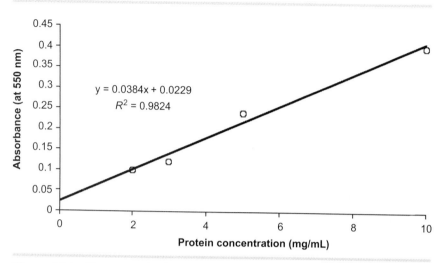

Figure A2.16 Biuret standard curve with trendline extrapolated backwards to the origin and the equation of the line displayed along with the R^2 value

8. Depending on the sensitivity range of the assay, it may be appropriate to extrapolate the linear trendline backwards or forwards. To do this, on the **Options** tab, set **Forecast Backward** (or **Forward**) the appropriate number of units. In Figure A2.16, "2" was entered for **Forecast Backward**.

9. Select **OK**. The resulting figure should look like Figure A2.16. *Note*: If desired, single-click the equation box to change y and x in the equation to the actual symbols for the variables.

10. **To modify an existing trendline**, right-click the trendline.

11. Follow the instructions given in the section "Importing Graphs into Microsoft Word" to copy the graph into your text document.

Multiple Lines on One Set of Axes

In some experiments it is desirable to compare the results from a number of different treatments. Plotting the data on one set of axes is often the most efficient way to convey this information. How many lines should you put on one set of axes? The CBE Manual recommends no more than 8, but use common sense. You should be able to follow each line individually, and the graph should not look cluttered.

Because there will be a multitude of lines on the figure, it is critical to identify the individual sets of data by means of labels or a legend. Excel will generate a legend for each graph, provided you enter the appropriate titles in the spreadsheet. This can be accomplished by entering a short title in the first row of each of the data columns. These titles will be used by Excel to generate the legend.

Let's say you monitored the production of an enzyme called beta-galactosidase by *E. coli* grown under different conditions. Every 10 min for 1 hr, you measured the absorbance (which represents the concentration of beta-galactosidase) in each culture condition. Time is the independent variable that will be plotted on the *x*-axis. According to Excel convention, therefore, time should be entered in Column A. The next five columns will contain the absorbance data at each time for the five different conditions (Table A2.9).

1. Enter the data in the spreadsheet. Select all of the entries (column headings as well as numbers) and click the **Chart Wizard** button.

2. In Chart Wizard Step 1 of 4, Chart Sub-type, select **Scatter with data points connected** by smoothed or straight lines.

3. In Chart Wizard Step 2 of 4, make sure the data series is in **Columns**.

4. In Chart Wizard Step 3 of 4, make modifications to the following five tabs:

   🗀 **Titles**

   Chart Title: **Leave blank** if this graph is for a lab report or poster.

TABLE A2.9 Sample data for beta-galactosidase production in *E. coli* grown under five different conditions

	A	B	C	D	E	F
1	Time (min)	Mutant, Gly	WT, Gly	WT, Gly IPTG	WT, Glu	WT, Glu IPTG
2	0	2	0.205	0.1	0.04	0.044
3	10	2	0.265	0.88	0.095	0.174
4	20	2	0.406	1.4	0.2	0.214
5	30	2	0.351	2	0.23	0.3
6	40	2	0.386	2	0.33	0.442
7	50	2	0.56	2	0.35	0.468
8	60	2	0.67	2	0.42	0.91

Value (X) axis: **Enter the x-axis label with the units in parentheses.** In the current example: Time (min).

Value (Y) axis: **Enter the y-axis label with the units in parentheses.** In the current example: Absorbance (at 420 nm).

🗀 **Axes**

No changes.

🗀 **Gridlines**

There should be no gridlines on the figure. Clear the box next to Value (Y) axis: Major gridlines.

🗀 **Legend**

A legend (key) is required when there is more than one line on the graph. The box next to Show legend should be checked.

🗀 **Data labels**

None.

5. With so many lines on one set of axes, it is a good idea to **enlarge the graph** while still in Excel. Pull on a corner of the frame to make this adjustment.

6. **Change the plot area to white, remove the borders, change the symbols (markers), and adjust the scale of the axes** as described in Table A2.7.

7. Double-click the legend and **remove the border.** Drag the legend inside the axes.

8. The final graph should look something like Figure A2.17.

Multiple Trendlines on One Set of Axes

This is similar to the previous example, except that the lines are trendlines, instead of smoothed or straight lines connecting the data points.

Table A2.10 shows sample data for an experiment in which enzyme (catalase) activity was measured for different combinations of inhibitor (hydroxylamine) and substrate (hydrogen peroxide) concentrations. Column A gives the values for the x-axis ($1/[H_2O_2]$ (M)), while Columns B–E give the data points for the y-axis ($1/$units of product formed $\cdot$ sec^{-1}). Each column (line on the graph) represents *one* inhibitor concentration, which you enter in Row 1 for Columns B–E. These short titles are used by Excel to generate the legend.

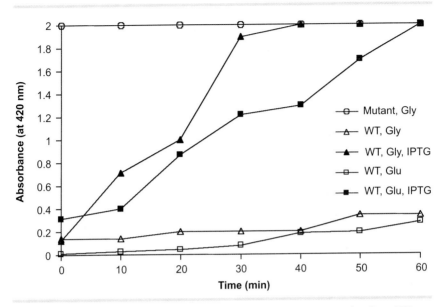

Figure A2.17 Beta-galactosidase production in *E. coli* grown under five differ-ent conditions. The mutant strain, which produces a non-functional repressor protein, was grown in glycerol. The wild-type strain was grown in glycerol or glucose, with or without the addition of IPTG (a lactose analog) at *t* = 0 min.

1. Enter the data in the spreadsheet. *Note:* The column titles must begin with I =, otherwise Excel confuses the number and per-cent symbol for a formula. Select all of the entries (column headings as well as numbers) and click **Chart Wizard**.

2. In Chart Wizard Step 1 of 4, Chart Sub-type, select **Scatter without connecting the points**.

3. In Chart Wizard Step 2 of 4, make sure the data series is in **Columns**.

4. In Chart Wizard Step 3 of 4, make modifications to the follow-ing five tabs:

   📁 **Titles**

   Chart Title: **Leave blank** if this graph is for a lab report or poster.

   Value (*X*) axis: **Enter the x-axis label (unformatted) with the units in parentheses**. In the current example: 1/[H2O2] (M).

TABLE A2.10 Effect of hydroxylamine concentration on catalase activity for different concentrations of hydrogen peroxide

	A	B	C	D	E
1	1/[H2O2] (M)	I = 0%	I = 0.0025%	I = 0.025%	I = 0.25%
2	7.14	66.3	60.74	93.34	130.17
3	3.45	39.66	31.3	72.45	114.77
4	0.74	33.77	32.97	48.36	98.76
5	0.34	13.63	14.99	16.81	31.02

Value (Y) axis: **Enter the y-axis label (unformatted) with the units in parentheses**. In the current example: 1/Catalase activity (units of product formed sec–1).

📁 **Axes**

No changes.

📁 **Gridlines**

There should be no gridlines on the figure. Clear the box next to Value (Y) axis: Major gridlines.

📁 **Legend**

A legend (key) is required when there is more than one line on the graph. The box next to Show legend should be checked.

📁 **Data labels**

None.

5. With so many lines on one set of axes, it is a good idea to **enlarge the graph** while still in Excel. Pull on a corner of the frame to make this adjustment.

6. **Change the plot area to white, remove the borders, change the symbols (markers), adjust the scale of the axes, and format the axis labels** as described in Table A2.7.

7. Double-click the legend and **remove the border**.

8. Now we want to have Excel draw a linear trendline for each of the data sets. Click one of the symbols to select the data set. All the data points in that set should now be highlighted (yellow).

9. Select **Chart | Add Trendline** on the menu bar.

10. There are two tabs under the Add Trendline option: **Type** and **Options**. On the **Type** tab, select the default (**Linear**).

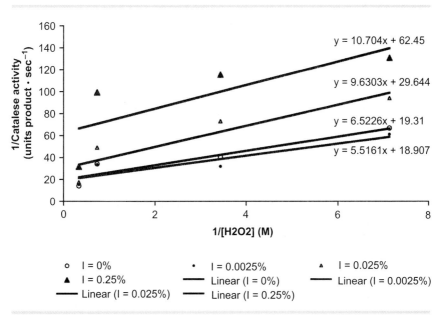

Figure A2.18 Effect of hydroxylamine concentration on catalase activity at different substrate concentrations. See Figure A2.19 for final form.

11. Now click the **Options** tab. Click the box next to **Display equation on chart**. The equations allow you to determine whether all the lines have a common y-intercept, a common x-intercept or neither. This information is used to determine whether hydroxylamine is a competitive or noncompetitive inhibitor in this reaction.

12. Select **OK**. The resulting figure will look something like Figure A2.18.

13. Since the "Linear" entries in the legend are meaningless, **single-click each of these text boxes** (to activate the frame), and then press the **Delete** key. Drag the legend inside the axes.

14. The resulting final graph will then look like Figure A2.19.

Figures as an Appendix to the Lab Report

Whenever possible, figures should be inserted in the document as close to their textual reference as possible. In some instances, however, you may be required to attach the figures on separate pages at the end of the report. To make the figure caption in that case:

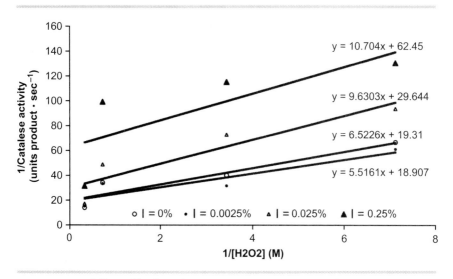

Figure A2.19 Effect of hydroxylamine concentration on catalase activity at different substrate concentrations

1. Follow the instructions given in the sections "Entering Data in Spreadsheet" and "Using Chart Wizard to Plot the Data" to make the first draft of the graph.

2. Modify the graph according to CBE Manual specifications (see Table A2.7).

3. When the graph is in its final form, single-click the Chart Area (the area inside the frame but outside of the axes).

4. From the menu bar, select **File | Print Preview | Set-up**.

5. Select the **Header/Footer** tab, and click the **Custom Footer** box. In the Left section of the footer area (which allows a maximum of 255 characters), type the word "Figure" followed by one space and then the appropriate figure number, followed by two spaces and then the figure title. Select the **A** box to change the font (Arial, regular, 12 pt is acceptable). Select **OK** to close all of the **Set-up** windows. The figure caption will be displayed in the proper position in the **Print Preview** window.

6. With the Chart Area still selected, click **File | Print**.

Importing Tables into a Word Document

Tables are handy for recording large amounts of data. Raw data should not, however, be included in your lab report or scientific paper. Furthermore, do not include both a table and a figure when they contain the same data.

Tables are preferable to figures in the following situations:

- To show precise numeric information rather than just the trend (as conveyed by a figure)
- To summarize information
- To describe information that is too complex to be shown in any other form

Because the Excel worksheet is really just one huge table, it may be easier to enter data in Excel, and then import the table into the Word document. Follow these steps:

1. Construct a well-organized table in Excel. Arrange the dependent variable data in columns with descriptive headings, so that like data are compared in columns, rather than across rows. If the variables presented have units, place the units in parentheses after the title in the column heading.
2. Use the mouse to select the data you wish to copy.
3. Click the **Copy** button on the Standard toolbar (or **Edit | Copy** on the menu bar).
4. Position the cursor in the Word document at the location where you wish to insert the table.
5. Click the **Paste** button on the Standard toolbar (or **Edit | Paste** on the menu bar).
6. Once the table is in the Word document, you can modify the table format (see Appendix 1, Section 2.12).

Preparing Oral Presentations with Microsoft PowerPoint

Microsoft PowerPoint allows you to create a slide show containing text, graphics, and even animations. The menus in PowerPoint parallel those in Microsoft Word, so if you are already familiar with Word, learning PowerPoint should not be difficult.

The screen drops in this appendix are for PowerPoint 2000, but the instructions also apply to PowerPoint XP, unless otherwise noted. Path notation for reaching a particular command or option is shown relative to a command on the menu bar, toolbar, or task bar. For example, **View | Toolbars | Drawing** means "Click the View button (on the menu bar), then select Toolbars, and then click Drawing."

Starting PowerPoint

To start PowerPoint, click the **PowerPoint** button on the taskbar at the bottom of your screen. Alternatively, click **Start | All Programs | Microsoft Office | Microsoft PowerPoint**. In PowerPoint 2000, a dialog window asks if you want to create a new presentation or open an existing presentation. If you are making a presentation in PowerPoint for the first time, click **Create a New Presentation Using Blank Presentation**. If you've already made a presentation and want to revise it, click **Open an Existing Presentation | More Files...** to browse your computer for the appropriate file, or choose one of the recently used files from the list.

When you start PowerPoint XP, a title slide for a new presentation is automatically displayed. If you want to revise an existing presentation, use **File | Open** or locate the file on the Task pane on the right of the screen.

Choosing a Slide Layout

In PowerPoint 2000, when you click **Blank presentation**, the **New Slide** dialog window asks you to choose an **AutoLayout** (Figure A3.1). These layouts range from text only to combinations of text and visual aid (table, graph, or image) to visual aid only.

When you select an **AutoLayout** and then click **OK**, the AutoLayout appears in Normal view (Figure A3.2A). The Normal view in Power-Point 2000 shows a screen split into three panes:

- Outline of your presentation on the left
- Current slide on the right
- Speaker notes below the current slide

In PowerPoint XP, there is an additional pane, the Task pane (Figure A3.2B). The Task pane appears when you choose certain commands or when you click **View | Task Pane** on the menu bar. Follow the instructions in the placeholders in the current slide to enter text (Figure A3.3). The same text will appear in the outline to the left so that later you can rearrange the text or the slides if desired (see the section on "Revising and Polishing Presentations").

Sample Presentation

The sample PowerPoint presentation we are using here is about a laboratory exercise in which a novel recombinant DNA, made of a plasmid (pGEM) and a fragment of a virus that infects monkeys (SV40), was analyzed by agarose gel electrophoresis. Six different recombinant DNAs are

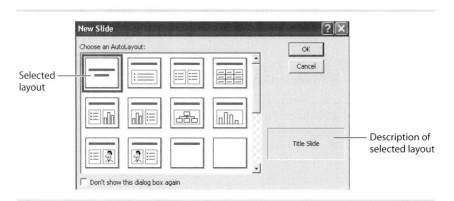

Figure A3.1 AutoLayouts for slides in PowerPoint 2000

(A)

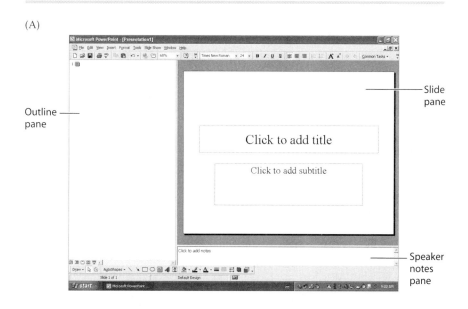

(B)

Figure A3.2 (A) A new slide in PowerPoint 2000 in Normal view. (B) The same slide in PowerPoint XP.

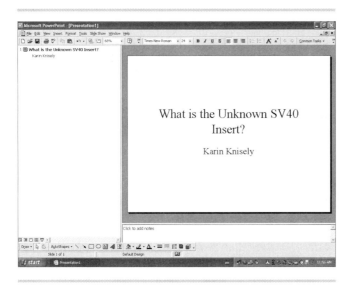

Figure A3.3 Text typed in the placeholders of the slide also appears in the outline to the left

possible, each one with the same plasmid (pGEM), but with a different SV40 insert. Since the SV40 inserts are all different sizes, and since agarose gel electrophoresis separates DNA fragments based on size, the goal of the exercise was to determine which of the six possible SV40 inserts was contained in the unknown recombinant DNA.

pGEM and SV40 were joined using a restriction enzyme called *Hind*III. In the lab exercise, the recombinant DNA was digested (cut apart) with *Hind*III, separating the DNA into its two components. The sample was then run on a gel, with the SV40 fragment migrating farther than the pGEM fragment because of its smaller size. By comparing the distance migrated by the SV40 fragment with a standard marker DNA, the identity of the SV40 fragment (based on its size) can be determined.

Creating a Title

To insert a title slide, click the **Title Slide Autolayout**. If this presentation were a lab report, a possible title might be "Analysis of recombinant DNA consisting of pGEM and SV40 using agarose gel electrophoresis." For an oral presentation to students who have all done the lab exercise, however, this long, descriptive title is not likely to generate excitement. **Know your audience and capture their interest immediately with a short, catchy title** such as the one in Figure A3.3.

Adding New Slides

Presentations in biology usually follow the Introduction-Body-Closing format. In the **Introduction**, the speaker guides the audience from information that is generally familiar to them to new information, which is the focus (**Body**) of the talk. The take-home message for the audience is given in the **Closing**.

To make the next slide, click **Insert | New Slide** from the menu at the top of the screen. Choose a slide layout from the **New Slide** dialog box (PowerPoint 2000) or from the Task pane (PowerPoint XP). A bulleted list works well if you want to give an overview of your talk first (Figure A3.4A). Each bulleted item should be a short phrase with keywords or key points, not full sentences. The phrases themselves should be informative but interesting for your audience. A good rule of thumb is to use no more than six words per bullet and no more than six bulleted items on one slide.

To delete a slide, display the slide you want to delete, and click **Edit | Delete Slide**.

Formatting Text

Just as in Microsoft Word, you can apply formatting such as font, font size, bold, italics, underline, and alignment **to selected words or lines in a single slide** by using the buttons on the Formatting toolbar. Select the text to be formatted by holding down the left mouse button and dragging across the text block from the first character to the last. Release the mouse button and click the desired formatting button on the toolbar.

To **subscript or superscript text**, select the character(s) to be formatted, click **Format | Font** on the menu bar, and click the appropriate box under **Effects** in the **Font** dialog box. This dialog box also allows you to change **text color**.

A good rule of thumb is to use **no more than two fonts or colors**, and to use them only for emphasis. For example, boldfaced, italicized, larger sized, and colored text stands out against the default text.

To **insert symbols** (Greek letters, mathematical symbols, Wingdings, etc.), click **Insert | Symbol** on the menu bar, and look for the desired symbol. For a uniform look, select the same font in the Symbol dialog window as the text font you are using in your PowerPoint presentation. It is not possible to make shortcut keys for symbols in PowerPoint, but if the symbols occur frequently in your presentation, you can copy and paste them to save yourself some keystrokes.

The **AutoCorrect** feature in Microsoft Word is also available in Power-Point (click **Tools | AutoCorrect**). Not only is it useful for correcting spelling mistakes, it can be used to replace long chemical names with a

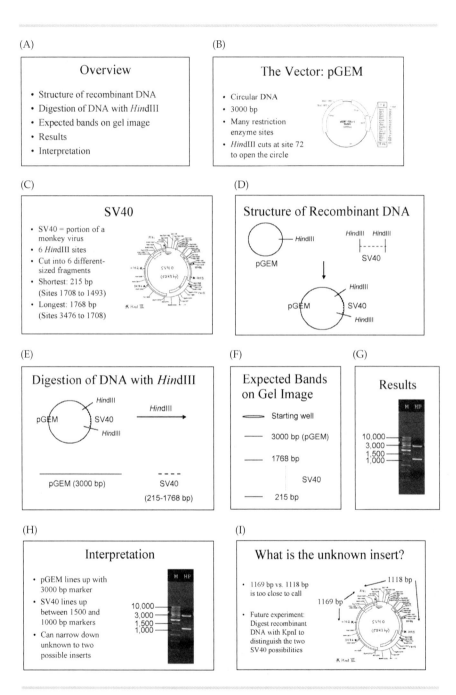

Figure A3.4 Slides with different combinations of text and visuals in a presentation about recombinant DNA analysis using agarose gel electrophoresis

simple keystroke combination and to italicize scientific names of organisms automatically. See Appendix 1, Section 4.6 for instructions on using the AutoCorrect feature.

If you want to make changes that **affect the text in every slide in the presentation**, see the section on "Slide Master and Title Master." The default font size (44 pt for titles and 32 pt for text) is designed to be large enough to be read even in the back of an average-sized classroom, so don't reduce font size to get more information on the slide. Instead, keep the wording simple.

Adding Images

A presentation with only text slides rarely gets your message across as effectively as one that combines text and visual aids. Visual aids take time to make or find, but they are worth the effort to capture and maintain audience interest.

The basic slide layout options are:

- Text only
- Text and a visual
- Visual only

While PowerPoint offers you many more than three slide layouts to choose from, you will find that most of the layouts are redundant. You can actually insert a visual into any placeholder in any of the slide layouts, but you can only type text in layouts that have text boxes. **To delete the existing placeholder box**, simply *right click* the placeholder and select **Cut**.

Figures A3.4B, C, H, and I are slides from the sample presentation, which have **both text and a visual**. Figures A3.4D, E, F, and G are slides with **just a visual**, and Figure A3.3 (the title slide) and Figure A3.4A are **text-only** slides.

Follow the instructions in the text placeholders to add a title or text. To insert a visual, you have several options.

Copy and Paste

Click the image from a Word, Excel, or other file and then **Edit | Copy** (or use the Copy button (📋) on the Standard toolbar, keyboard shortcut Ctrl+C, or right mouse button menu to copy). Click the desired location in the PowerPoint slide and click **Edit | Paste**, or use the **Paste** button (📋), keyboard shortcut Ctrl+V, or right mouse button to paste. Make sure the text (if any) in the visual is large and legible (in Figure A3.4B and C, the text on the actual slide meets these criteria). To resize the image:

- Click the image to display the selection handles around the frame
- Position the mouse pointer over a handle in one of the corners to display a two-headed arrow
- Hold down the left mouse button and drag the frame to the desired size, and then release the mouse button

To reposition the image, position the cursor over the image to display crossed arrows, hold down the left mouse button, drag the image to the desired location, and release the mouse button.

Insert | Picture

Click **Insert | Picture** from the menu bar, and then click the source of the picture (Clip Art, From File, etc.). PowerPoint readily accepts .jpg, .gif, .bmp and other standard image file formats. The gel image in Figure A3.4G and H was added to the slide using the **Insert | Picture** command. **Resize and reposition if necessary** (see the section on "Copy and Paste"). To crop the picture, select it, and then click **Format | Picture** from the menu bar. You can crop the picture precisely by entering numbers for Left, Right, Top, and Bottom on the **Picture** tab.

Drawing Toolbar

To add simple graphics to your slide, click **View | Toolbars | Drawing** to display the Drawing toolbar at the bottom of the screen (Figure A3.5). **To add a line**, for example, click the line on the Drawing toolbar to display crosshairs. Position the mouse pointer on the slide where you want the line to start, hold down the left mouse button, drag to where you want the line to end, and release the mouse button.

To change the angle or length of the line, click the line and then position the mouse pointer over one of the end points of the line to display a two-headed arrow. Hold down the left mouse button, drag to where you want the line to end, and release the mouse button.

Drawing toolbar

Figure A3.5 Drawing toolbar for making simple graphics

To change the location of the line, click the line and then position the mouse pointer over the line to display crossed arrows. Hold down the left mouse button, drag the line to the desired location, and release the mouse button. For more precise positioning, click the line to display the selection handles. From the menu bar, select **Format | AutoShape | Layout** and click the **Advanced** button. Change the numbers to place the line exactly where you want it.

To format the line, click the line to display the selection handles. From the menu bar, click **Format | AutoShape | Colors and Lines** to change the appearance of the line or **Format | AutoShape | Size** to size it precisely.

To add text to your drawing, click the **Text Box** button on the Drawing toolbar at the bottom of the screen. Position the mouse pointer where you want the text box to appear, hold down the left mouse button, drag to enlarge the text box to the size you think you'll need (you can resize it later), and release the mouse button. Now you can type the text inside the box. Font and font size can be changed by selecting the text and using the commands on the menu bar or toolbars. To format the text box itself, click the text box to display the selection handles on the frame. Click **Format | Text Box** to open a dialog box that allows you to change the appearance, size, position, and word wrapping of the text box.

If individual objects make up a graphic, then it may make sense to group them as a unit. Grouping allows you to copy, move, or format all of the objects in the graphic at one time. **To group individual objects into one unit**, click each object while holding down the **Shift** key. Release the **Shift** key, click the **Draw** button on the Drawing toolbar, and then click **Group**. One set of selection handles surrounds the entire unit when the individual objects are grouped. To ungroup, simply select the group, click the **Draw** button, and click **Ungroup** (Figure A3.6). For an example of line drawings that were created using the Drawing toolbar, see Figures A3.4D–F.

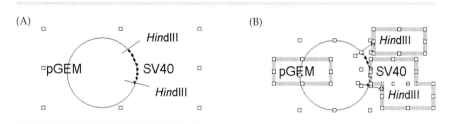

Figure A3.6 Individual graphics grouped (A) and ungrouped (B)

Adding a Table

To make a table that spans the entire slide, click **AutoLayout | Table** (PowerPoint 2000) or select a layout with one column from the task pane (PowerPoint XP). When you double-click the placeholder in **AutoLayout | Table**, a dialog box asks you to choose the number of columns and rows for the table. To insert a table into a one-column slide, click **Insert | Table** and enter the desired number of columns and rows. A table then appears in the slide, and all of the cells are the same size. After typing text in any of the cells, you can change the size of the cells manually. To do this, position the mouse pointer on one of the horizontal ($\doteq$) or vertical lines ($\leftrightarrow$). Hold down the left mouse button, move the pointer to adjust the row or column size, and then release the mouse button. There are other formatting options located on the Tables and Borders floating toolbar, such as splitting and merging cells and inserting or deleting columns or rows.

Small tables can also be added to slides with two-column format. Click one of the placeholders and select **Insert | Table** from the menu bar. Enter the number of columns and rows, and click **OK**.

Adding a Graph

If you are already familiar with making graphs in Microsoft Excel, then the easiest way to add graphs to a PowerPoint presentation is to copy and paste as follows:

- Make a graph in Excel (see Appendix 2).
- Single-click the Chart Area to activate the graph. Click **Copy**.
- In PowerPoint, click **Insert | New Slide** and choose a layout suitable for the graph.
- Click **Paste**. Resize and reposition the graph as needed (see the section on "Copy and Paste").

The Chart AutoLayout (PowerPoint 2000) and the Title and Chart layout (PowerPoint XP) may seem like logical choices for adding a chart (graph), but the default Microsoft Graph program associated with these slides is not as versatile as Excel for making graphs according to scientific conventions.

Navigating among Slides

There are several ways to move from one slide to another when preparing a presentation: In **Normal view** (**View | Normal**; Figure A3.7):

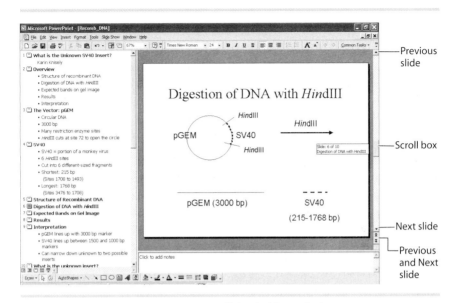

Figure A3.7 Ways to navigate among slides in Normal view

- Click the Page Up or Page Down key on the keyboard
- Use the vertical scroll bar on the right-hand side of the Slide pane
- Inside the scroll bar, drag the scroll box up or down to move backwards or forwards through the presentation.
- Click the **Previous Slide** or **Next Slide** arrow buttons to move from slide to slide.
- On the left-hand side of the screen is the Outline pane. Click the text on the slide you want to display. *Note*: In PowerPoint XP, there are two tabs on the Outline pane: **Outline** (text) and **Slides** (thumbnails). Choose the view you prefer.

In **Slide Sorter view** (**View | Slide Sorter**), single-click the desired slide to select it. Double-click it to edit the contents in Normal view.

Saving and Printing Presentations

To save a presentation for the first time, click **File | Save as** on the menu bar. Do not wait until you are finished making the entire presentation to save the file. Save it after you have made the first slide. Choose a meaningful name for the file; PowerPoint automatically saves the file as a "Presentation," with the .ppt extension. At the top of the dialog box, click the down arrow next to the **Save in** text box to select the location (drive, folder, sub-folder) where you want to save the file.

You can print all of the slides in a presentation or just selected slides. You can also print the outline, handouts, and the speaker notes.

When printing the **Slides**, first specify the desired size and orientation in the **File | Page setup** dialog box. The default settings are fine if you decide to make overheads for your presentation (e.g., if the room where you will be giving your talk does not have a computer and projection equipment for a PowerPoint presentation). Then click **File | Print** from the menu bar and make the appropriate selections there. When printing on a black-and-white printer, select Grayscale to avoid printing out the background of the slides (see the section on "Special Effects").

The **Handouts** option is useful for printing 1, 2, 3, 4, 6, or 9 slides on one page. If you want the audience to take notes on the handouts, print out fewer slides per page. Select the orientation of the page (portrait or landscape) in **File | Page setup**. Select the order of the slides (across or down the page) in **File | Print**.

The **Notes Pages** option allows you to print out the slide and the speaker notes below it. The speaker notes are comparable to notecards in that they help you remember the important points to make about each slide. By sticking to the script, you are likely to stay within your time limit and avoid getting sidetracked. As with Handouts, the orientation of the Notes pages is selected in **File | Page setup**. To print out two or more Notes pages on one sheet of paper, click **File | Print | Properties** and choose the desired number of **Pages per Sheet**.

Revising and Polishing Presentations

PowerPoint provides **multiple views** for displaying a presentation: Normal, Slide Sorter, Notes Page, and Slide Show. You can switch views two ways:

- Click the **View** command on the menu bar at the top of the screen and select a different view
- Click one of the buttons in the lower left corner of the screen

Slide Sorter view is useful for evaluating the overall appearance of a presentation, because it displays all of the slides in a presentation in miniature, complete with formatting (see the section on "Special Effects"). You cannot edit the contents of an individual slide in Slide Sorter view, but you can rearrange the slides.

Click **View** | **Slide Sorter** to change to Slide Sorter view. To select a slide to move, copy, add, or delete, single-click it. To edit the contents of the slide in Normal view, double-click it.

Moving Slides

To move a slide, position the mouse pointer on the slide you want to move, hold down the left mouse button and drag the vertical line to a new location between slides, and release the mouse button. The slide now appears at its new position. You can also move a slide using **Edit** | **Cut**, clicking at the new insertion point, and clicking **Edit** | **Paste**.

To move a group of adjacent slides, click the first slide in the group, hold down **Shift** on the keyboard, and click the last slide in the group. All slides will be marked with a highlighted border. Hold down the left mouse button, drag the vertical line to a new insertion point between slides, and release the mouse button. The group of slides appears at its new position. As with a single slide, you can also use the **Cut** and **Paste** commands on the menu bar.

To move a group of nonadjacent slides, follow the procedure for adjacent slides, but hold down **Ctrl** instead of **Shift** on the keyboard. The slides appear in the same consecutive order at the new location.

Adding and Deleting Slides

To add a new slide, click the slide that you want the new slide to follow. Then click **Insert** | **New Slide** and choose a slide layout. Click **OK** and a new slide is inserted in the presentation. Double-click the new slide to add content.

To delete an unwanted slide, single-click the slide and press **Delete** on the keyboard. To correct the deletion, click **Edit** | **Undo Delete Slide**.

Copying Slides

To copy a slide, click the slide you want to copy and then **Edit** | **Copy**. Click the insertion point between slides and then click **Edit** | **Paste**. You can also copy a slide from another presentation using this method. If you want the duplicate slide to immediately follow the original, click the slide to be copied and use **Edit** | **Duplicate**. The duplicate will be inserted automatically after the original. To change the layout on the duplicate, simply single-click the duplicate slide, click **Format** | **Slide Layout**, and select a new layout for the slide. This method was used to make Figure A3.4H from Figure A3.4G.

Slides can also be rearranged in the Outline pane in Normal view using the same procedures just described.

Spell Check

Don't let an otherwise interesting and well-organized presentation be ruined by typos on the slides. As with Microsoft Word, PowerPoint offers three ways to correct spelling and grammar:

- AutoCorrect
- Automatic spelling and grammar checker
- Manual, systematic spelling and grammar check

See Appendix 1, Section 3.4 for instructions on using these spell checkers.

Special Effects

While your presentation will be evaluated primarily on content and organization, appearance definitely affects your instructor's state of mind. Table A3.1 gives some recommendations for the appearance of your slide show.

There is nothing wrong with a well-rehearsed black-and-white slide show. On the other hand, we are so accustomed to colorful animations and interactive computer programs that we have come to expect this level of entertainment in all aspects of our daily lives. While scientists view themselves primarily as conveyors of information rather than

TABLE A3.1 Presentation tips

DO	DON'T
Keep the wording simple	Write every word you're going to say on the slide
Make the text large and legible	Vary the background or use distracting slide transitions and sound effects
Strive for a consistent look (use the same font and format for each slide)	Include tables when you can show the trend better with a graph
Include visual aids to provide both visual and auditory information	Talk endlessly without referring to a visual
Allow enough time for each slide	Rush through the slides without mentioning the take-home message about each one

entertainers, adding some well-chosen special effects to our presentations may help get the message across more effectively.

Background. An easy way to change the background of your slides is to add color or apply a design template to an existing presentation. Open your presentation (see the section on "Starting PowerPoint") and click **View | Slide Sorter**. This view allows you to see how changes affect the entire slide show.

PowerPoint offers a wide variety of design templates that give your slides a uniform and professional look. Click **Format | Apply Design Template** (in XP, **Format | Slide Design**) to view some of them. Templates with a light background make the room much brighter during the actual presentation, which has the dual advantage of keeping the audience awake and allowing you to see your listener's faces. Many professional speakers, however, prefer white text on a dark background because the text appears larger. Whatever your preference, choose a template that reflects your style, is appropriate for the topic, and complements the content of the individual slides.

In PowerPoint 2000, **to restore the default template** after having selected a design template, click **Format | Slide Color Scheme** and then the **Custom** tab. Double-click **Background** and change the color to white. Do the same for **Shadows** and **Fills**. Then double-click **Text and lines** and **Title text** and change their color to black. Click **Apply to All**. Finally, click **Format | Background** and check the box next to **Omit background graphics from master**.

In PowerPoint XP, it's much easier to restore the default template. Click **Format | Slide Design** on the menu bar or the **Design** button (with a pencil on it) on the Standard toolbar to display the Slide Design task pane. Under Available for Use, click the first item, **Default Design**.

Still in Slide Sorter view, click **Format | Background.** Click the down arrow to select a different background color for your slides. You may apply the color to the currently selected slide or to all slides, but a consistent format is preferred.

Slide Master and Title Master. The slide master controls the **appearance of the text and objects** in every slide except the title slide. The title master allows you to apply a format to the title slide, which is different from that in the rest of the slides.

Note: If you have not applied a design template to your presentation, the title master command is unavailable and changes made in the slide master also affect the title slide.

Click to edit Master title style

Title Area for AutoLayouts

- Click to edit Master text styles
  - Second level
    - Third level
      - Fourth level
        - » Fifth level

Object Area for AutoLayouts

⟨date/time⟩	⟨footer⟩	⟨#⟩
Date Area	Footer Area	Number Area

Figure A3.8 Default slide master (no design template is applied)

You can modify the slide master and title master at any time—before, during, or after making the first draft of your presentation. Click **View | Master | Slide Master** to see the default design (Figure A3.8) or the currently applied design template. The title master is located in the same submenu as the slide master and contains a Subtitle Area instead of an Object Area.

To change the font, font style, font size, or font color, click a placeholder in the Title Area or Object Area and select **Format | Font** from the menu bar at the top of the screen. To change the bullet, click anywhere on the bulleted line and select **Format | Bullets and Numbering**. You can also insert objects such as clip art or images into the Title Area or Object Area. Click **Insert | Picture** and then make the appropriate selection. **The changes affect every slide in the presentation, except the title slide** (see exception in the note above).

The color scheme of the slide master is determined by the selected design template to ensure that the text color harmonizes with, and is legible against, the background color. **When you change the slide master, however, you override the default color scheme**. In other words, the text colors you specify in the slide master are applied instead of the color scheme's text colors. Strive to maintain good contrast whenever you modify the text and background colors.

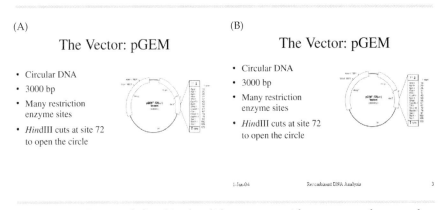

Figure A3.9 Footers defined in the slide master or title master can be turned off (A) or on (B) on the Slide tab in the Header and Footer dialog box.

Adding Headers and Footers. The three boxes at the bottom of the master allow you to enter the date and time, a custom footer, and the page number. While it is not customary to display this information during a slide show, it may be handy to have it on the printed slides if they are used as handouts.

To turn off footer options for a slide show (Figure A3.9A):

- Click **View | Header and Footer** on the menu bar to display the **Header and Footer** dialog box.

- On the **Slide** tab, clear the Date and Time, Slide Number, and Footer check boxes.

Even if you've added a date, custom footer, or page number to the slide master or title master, these will not be displayed on the slide.

To display footers defined in the slide master or title master on the printed slide itself (Figure A3.9B):

- Click **View | Header and Footer** on the menu bar to display the Header and Footer dialog box.

- On the **Slide** tab, select the Date and Time, Slide Number, and Footer check boxes.

- Click **Apply to All** to print this information on every slide.

- Click **File | Print** and select **Slides** from the **Print What** drop-down list.

- Make any other selections in the **Print** dialog box and then click **OK**.

Footers defined in the slide master or title master only appear on the slide (if you have made the appropriate selections), not in the footer of a Handouts or Notes page (to print out Handouts or Notes pages, click **File | Print | Print What**). You may want to define a header and footer on these printed materials, which is different from the footer you have on the slides themselves.

To define a different header and footer, follow these steps:

- Click **View | Header and Footer** from the menu bar to display the **Header and Footer** dialog box.

- On the **Slide** tab, clear the Date and Time, Slide Number, and Footer check boxes.

- On the **Notes and Handouts** tab, select the relevant check boxes (Figure A3.10). The default settings are date and time displayed in the upper right corner and the page number in the lower right corner. A custom header and footer can be entered in the respective boxes if desired.

- Click **Apply to All** to print the selected information on every page.

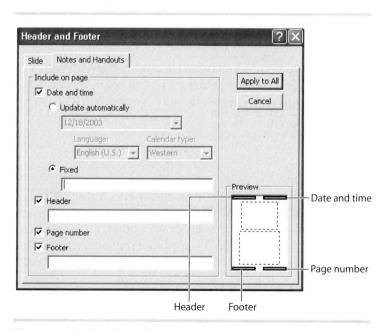

Figure A3.10 Header and Footer dialog box allows you to print date and time, page number, and a custom header and footer on Handouts and Notes pages

- Click **File** | **Print** and select Handouts or Notes Pages from the **Print What** drop-down list.
- Make any other selections in the **Print** dialog box and then click **OK**.

Slide Transitions. Transitions affect the way slides appear in a slide show. To see how your presentation looks without slide transitions, click the first slide in **Slide Sorter view** or display the first slide in Normal view and then change the view to Slide Show (using **View** | **Slide Show** or the corresponding button in the lower left corner of the screen). The first slide will stay on the screen until you click the left mouse button, press **Enter**, or press the space bar on the keyboard to advance to the next slide. The slides in the presentation are displayed sequentially unless you press the **Esc** key to end the show, click a button on the Slide Show toolbar in the lower left corner of the screen, or right click and choose an option from the menu.

To see what different transitions look like in PowerPoint 2000, switch to **Slide Sorter** view and locate the toolbar with the Slide Transition Effects and Preset Animation boxes (Figure A3.11A). In PowerPoint XP, click the **Transition** button on the right-hand side of the toolbar to display the Slide Transition Task pane (Figure A3.11B). **If you choose to add Slide Transition Effects to your presentation, make sure they do not detract from the content!**

To apply a basic transition effect to a slide, click the slide, and then select an effect from the drop-down menu in the **Slide Transition Effects** box (PowerPoint 2000) or the Slide Transition Task pane (PowerPoint XP). You will see an instant preview of the effect. If you missed it, click the small icon below the slide (Figure A3.12), the **Animation Preview** button on the Slide Sorter toolbar (PowerPoint 2000), or the **Play** button on the Task pane (PowerPoint XP) to replay the effect.

To apply a single effect to all the slides, click **Edit** | **Select All** and then choose an effect from the **Slide Transition Effects** drop-down menu (PowerPoint 2000). To clear the Select All option, click between any two slides or after the last slide. **To apply a single effect only to selected slides**, hold down **Shift** or **Ctrl** on the keyboard (see the section on "Moving Slides") and select the slides. In PowerPoint XP, you select the slides affected by the transition on the Task pane.

To modify the Slide Transition Effect selected for a slide, click the **Slide Transition** button (see Figure A3.11A). In the **Slide Transition** dialog box, you can change the effect, the speed for the effect, and how long the slide is displayed on the screen. You can even add sound to the

(A)

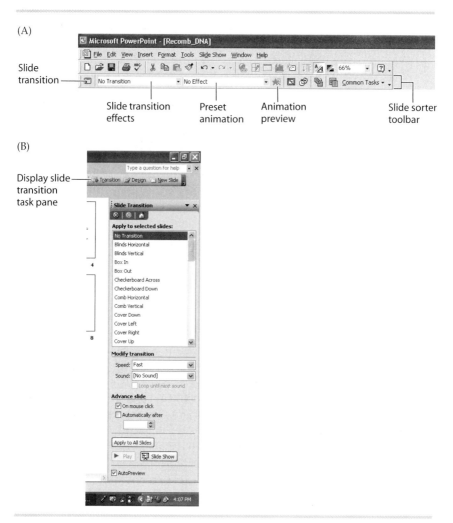

Slide transition

Slide transition effects

Preset animation

Animation preview

Slide sorter toolbar

(B)

Display slide transition task pane

Figure A3.11 (A) In PowerPoint 2000, the Slide Sorter toolbar allows you to select effects for slide transitions. (B) In PowerPoint XP, slide transitions are selected on the Task pane.

slide. These options are available on the Slide Transition Task pane in PowerPoint XP. After you have selected the desired options, click **Apply** to apply them to the selected slide(s) or **Apply to All** to apply them to all the slides in the presentation.

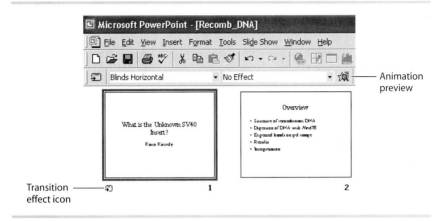

Figure A3.12 Transition effect selected for Slide 1 is "Blinds Horizontal." To view the effect, click the Transition effect icon or the Animation Preview button. In PowerPoint XP, these options are available on the Slide Transition Task pane.

Multimedia Files. Movie and video clips, music, and sounds may enhance your presentation. To insert multimedia files, first make sure you are not in Slide Sorter view. Then select **Insert | Movies and Sounds** from the menu bar and click the type. A dialog box asks if you want the clip to play automatically when you reach that slide in the slide show or only when you click it. After selecting **Yes** or **No**, an icon will be placed on the slide.

Not all movie and video clips can be imported into PowerPoint. Some require the installation of additional software such as QuickTime for the clip to work. If you are having problems playing movie or sound clips in PowerPoint, ask someone in computer services for help.

Delivering Your Presentation

A PowerPoint presentation can be "delivered" electronically by attaching it to an e-mail, but most presentations are delivered in person. Do your homework and find out what audio-visual equipment is available in the room where you will hold your talk. If you only have a chalkboard, just print out the speaker notes. If you'll have access to an overhead projector and screen, print out the slides and make transparencies. These printing options are explained in the section on "Saving and Printing Presentations."

Running the Show

To deliver a PowerPoint slide show, you'll need a computer and projection equipment as well as a screen on which to display the slides. If the presentation room does not have this equipment, ask your host if he/she can provide it or, if you are affiliated with a college or university, you may be able to borrow a laptop computer and projector for your presentation.

There are several ways to run a PowerPoint presentation:

- From a Web server
- From a disk or CD on a computer at the presentation site
- From your laptop computer

To run a presentation from a Web server, first find out if you have access to the Internet at the presentation site. If so, save your file as a Web page and write down the address of the Web folder. Similarly, if your college or university has networked computers, you can prepare your presentation on a networked computer in your room, for example, save the file in your Private Space, and then access the file from another networked computer in the room where you will hold your talk.

If the presentation site has a computer, but it is not hooked up to the Internet, save your .ppt file on a removable disk and carry it with you. Ask your host if PowerPoint is installed on the presentation room's computer. If so, just insert the disk or CD, start PowerPoint, open your .ppt file, and click the **Slide Show** button in the lower left corner of the screen.

If the presentation site's computer does not have PowerPoint installed on it, you may be able to install PowerPoint from a set of disks you bring with you. To learn more about this option, ask your system administrator.

If you have a laptop computer, make sure PowerPoint is installed on it, and then save your .ppt file on the hard drive. At the presentation site, start PowerPoint, open the .ppt file, and click the **Slide Show** button in the lower left corner of the screen to begin your presentation.

Navigating among Slides during a Slide Show

Click the **Slide Show** button in the lower left corner of the screen to begin your presentation. If you have not specified a time for the slides to advance automatically (see the section on "Slide Transitions"), the first slide stays on the screen until you click the left mouse button, or press **Enter** or the space bar on the keyboard. A cordless mouse allows you to change slides from anywhere in the room.

There are two ways to change slides with the mouse:

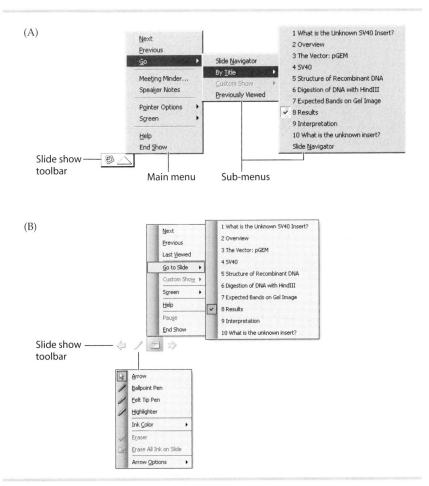

Figure A3.13 Options for navigating among slides during a slide show. (A) In PowerPoint 2000, clicking the Go command on the Slide Show toolbar opens further menus that allow you to select and display a specific slide. (B) In PowerPoint XP, the toolbar has arrows to go forward and backward, a pointer button, and a slide button for navigation options.

- Click a button on the Slide Show toolbar in the lower left corner of the screen (Figure A3.13).
- Click the right mouse button to display the main menu of navigation options

To end a slide show, press the **Esc** key on the keyboard or select **End Show** from the pop-up menu.

BIBLIOGRAPHY

Bregman A. 1996. Laboratory investigations in cell and molecular biology. Rev. 3rd ed. New York: John Wiley & Sons. 336 p.

Calabria J, Burke D. 1997. Microsoft Word 97 exam guide. Indianapolis: QUE Corp.

Corrections to Publications: Corrections to Scientific Style and Format (6th ed.) Updated June 2000 Available at <http://www. councilscienceeditors.org/ publications/ corrections6th.cfm> Accessed 2004 Oct 20.

D'Arcy J. 1998. Technically Speaking: A Guide for Communicating Complex Information. Columbus: Battelle Press. 270 p.

Fegert F, Hergenröder F, Mechelke G, Rosum K. 2002. Projektarbeit: Theorie und Praxis. Stuttgart: Landesinstitut für Erziehung und Unterricht Stuttgart. 226 p.

Glenn W. 2000. Word 2000 in a nutshell. Sebastopol, CA: O'Reilly & Associates, Inc. 491 p.

Hacker D. 1997. A pocket style manual. 2nd ed. Boston: Bedford Books. 185 p.

Hailman JP, Strier KB. 1997. Planning, proposing, and presenting science effectively. Cambridge: Cambridge University Press. 150 p.

Harnack A, Kleppinger E. 2001. Online! A reference guide to using internet sources. Boston: Bedford/St. Martins. 260 p.

Information Services and Resources, Bucknell University. 2004. Evaluating Websites as Sources. Available at <http://www.bucknell.edu/Library_computing/Doing_Research/Start_Your_Research_Here/Evaluating_Sources/Website.html> Accessed 2006 May 1.

Jones A, Reed R, Weyers J. 1994. Practical skills in biology. London: Longman Scientific & Technical. 292 p.

Lannon JM. 2000. Technical Communication, 8th ed. New York: Addison Wesley Longman, Inc. 680 p.

Light RJ. 2001. Making the most of college: Students speak their minds. Cambridge, MA: Harvard University Press. 242 p.

Lunsford A. 2002. The everyday writer. 2nd ed. Boston: Bedford/St. Martin's. 534 p.

Lunsford A, Connors R. 1999. Easy writer: A pocket guide. Boston: Bedford/St. Martin's. 266 p.

Lunsford A, Connors R. 1995. The new St. Martin's handbook. 3rd ed. Boston: Bedford/St. Martin's. 797 p.

McMillan VE. 1997. Writing papers in the biological sciences. 2nd ed. Boston: Bedford Books. 197 p.

Moore DS. 2000. The basic practice of statistics. 2nd ed. New York: W. H. Freeman and Company. 619 p.

Palmer-Stone D. Last revised July 31, 2001. How to Read University Texts or Journal Articles. Counselling Services, University of Victoria, Victoria, BC, Canada. Available at <http://www.coun.uvic. ca/learn/program/hndouts/Readtxt.html> Accessed 2004 Oct 20.

Pechenik JA. 2004. A Short Guide to Writing About Biology, 5th ed. New York: Pearson/Longman. 302 p.

Peterson SM. 1999. Council of Biology Editors guidelines No. 2: Editing science graphs. Reston, VA: Council of Biology Editors. 34 p.

Peterson SM, Eastwood S. 1999. Council of Biology Editors guidelines No. 1: Posters and poster sessions. Reston, VA: Council of Biology Editors. 15 p.

Sagman SW. 1999. Running Microsoft PowerPoint 2000. Redmond, WA: Microsoft Press. 553 p.

Samuels ML, Witmer JA. 1999. Statistics for the life sciences. 2nd ed. Upper Saddle River, NJ: Prentice Hall. 683 p.

Style Manual Committee, Council of Biology Editors. 1994. Scientific style and format: The CBE manual for authors, editors, and publishers. 6th ed. Cambridge, UK: Cambridge University Press. 825 p.

The University of Reading, IT Services. April 21, 2006. Equations and fields in Microsoft Word 2003 <http://www.rdg.ac.uk/ITS/info/training/notes/word/equations/> Accessed 2006 Aug 3.

Utts JM, Heckard RF. 2002. Mind on statistics. Pacific Grove, CA: Duxbury Press. 592 p.

VanAlstyne JS. 1986. Professional and technical writing strategies. Upper Saddle River, NJ: Prentice Hall, Inc. 433 p.

Vodopich D, Moore R. 1996. Biology laboratory manual to accompany biology. 4th ed. Boston: WCB/McGraw-Hill. 510 p.

Vogt C. 2004. Creating Equations with Microsoft Word. Information Systems & Technology, University of Waterloo, Waterloo, Ontario Canada. Available at <http://ist.uwaterloo.ca/ec/equations/equation.html> Accessed 2004 Aug 7.

Woolsey JD. 1989. Combating poster fatigue: How to use visual grammar and analysis to effect better visual communications. Trends in Neuroscience 12(9): 325–332.

Index

About the Book

Editors: Andrew D. Sinauer, Sara Tenney
Project Editors: Carol J. Wigg, Chelsea D. Holabird
Production Manager: Christopher Small
Cover Design: Jefferson Johnson
Book Design: Kristy Sprott/The Format Group LLC
Page Layout: Joanne Delphia